STANDARD METHODS OF

CLINICAL CHEMISTRY

VOLUME 5

CONTRIBUTORS TO THIS VOLUME

DANIEL H. BASINSKI · LOUIS BERGER · WENDELL T. CARAWAY · HELEN L. CHEUNG · HAROLD O. CONN WILLARD R. FAULKNER · DEAN C. FLETCHER · HOWARD S. FRIEDMAN · S. RAYMOND GAMBINO · ADRIAN HAINLINE, JR. HANS HOCH · FRANK A. IBBOTT · ROY B. JOHNSON, JR. ALEX KAPLAN · RODERICK P. MACDONALD · SAMUEL MEITES NATHAN RADIN · MIRIAM REINER · EUGENE W. RICE GUILFORD G. RUDOLPH · LEONARD T. SKEGGS, JR. · AGNES STUMPFF · RALPH E. THIERS · SEYMOUR WINSTEN BENNIE ZAK

EDITORIAL COMMITTEE

WILLARD R. FAULKNER, *Microchemistry Laboratory, Department of Pathology, The Cleveland Clinic Foundation, Cleveland, Ohio*

S. RAYMOND GAMBINO, *Director of Laboratories, The Englewood Hospital, Englewood, New Jersey*

RODERICK P. MACDONALD, *Director of Clinical Chemistry, Department of Laboratories, Harper Hospital, Detroit, Michigan*

EUGENE W. RICE, *Biochemistry Department, The William H. Singer Memorial Research Laboratory, Allegheny General Hospital, Pittsburgh, Pennsylvania*

STANDARD METHODS OF

CLINICAL CHEMISTRY

VOLUME 5

By the American Association of Clinical Chemists

Editor-in-Chief

SAMUEL MEITES

Biochemist, Clinical Chemistry Laboratory
The Children's Hospital
Assistant Professor, Department of Pediatrics
The Ohio State University College of Medicine

1965

ACADEMIC PRESS • New York and London

ACADEMIC PRESS, INC.
111 Fifth Avenue, New York, New York 10003

United Kingdom Edition published by
ACADEMIC PRESS, INC. (LONDON) LTD.
Berkeley Square House, London W.1

LIBRARY OF CONGRESS CATALOG CARD NUMBER: 53-7099

Second Printing, 1968

PRINTED IN THE UNITED STATES OF AMERICA

This volume is dedicated to Michael Somogyi (1883–   ).

MICHAEL SOMOGYI—A BIOGRAPHICAL SKETCH

When, in 1926, Michael Somogyi assumed the title of *Biochemist* at the Jewish Hospital of St. Louis, there was hardly a precedent for this position. A distinct discipline of biochemistry barely existed, in its modern sense, and the "early pioneers" were largely confined to posts at academic institutions. How unique it seems that a chemist, who had graduated in 1905 as a chemical engineer [sic!] and, in 1914, written a doctorial dissertation on catalytic hydrogenation, should take up a long and fruitful study (at a hospital) of the metabolism and physiology of carbohydrates, ketone bodies, and insulin, as well as of diabetes. His papers on the analysis of blood sugar and amylase remain classic as well as current for today's clinical chemist. Somogyi's interest in clinical biochemistry, however, is not accidental. After serving an assistantship in biological and pathological chemistry at the University of Budapest, Somogyi spent two years (1906–1908) assisting in the medical school at Cornell University. While there, he was a colleague of P. A. Shaffer who later (1922) invited him to join the staff of the Department of Biochemistry at Washington University, in St. Louis. Meanwhile (1908–1922), Somogyi (born in Reinersdorf, 1883) had returned to Budapest, obtained his doctorate (1914) at the University of Budapest, and worked for various municipal laboratories, while World War I and the Austrian-Hungarian Empire passed into history. After four years at Washington University (1922–1926), Dr. Somogyi toiled for three decades as the clinical chemist at the Jewish Hospital of St. Louis, before reaching the *emeritus* status in 1955. His research interests continue unabated. Within recent years Somogyi summarized one aspect of a lifetime of dedicated effort. A series of his articles pin-points the hazards of excess insulin administration to the adult diabetic. "Hypoglycemia begets hyperglycemia." Our awareness of this diabetogenic rebound effect, mediated by mobilization of insulin antagonists in response to hypoglycemia, stands as a legacy to his research in carbohydrate metabolism. Clinical chemists have honored Somogyi with the Ernst Bischoff Award (1953) and the Donald D. Van Slyke Award (1964). The creative efforts of Michael Somogyi, and the few like him, have supplied the very marrow to the growing science of clinical chemistry.

CONTRIBUTORS

Numbers in parentheses indicate the pages on which the authors' contributions begin.

DANIEL H. BASINSKI, *Henry Ford Hospital, Department of Laboratories, Detroit, Michigan* (137)

LOUIS BERGER, *Sigma Chemical Company, St. Louis, Missouri* (211)

WENDELL T. CARAWAY, *Flint Medical Laboratory, and Laboratories of McLaren General Hospital, and St. Joseph Hospital, Flint, Michigan* (19)

HELEN L. CHEUNG, *District of Columbia General Hospital, Washington, D.C.* (257)

HAROLD O. CONN, *Veterans Administration Hospital, West Haven, Connecticut, and Department of Internal Medicine, Yale University School of Medicine, New Haven, Connecticut* (43)

WILLARD R. FAULKNER, *The Cleveland Clinic Foundation, Cleveland, Ohio* (199)

DEAN C. FLETCHER, *University of Nevada Desert Research Institute, and Washoe Medical Center, Reno, Nevada* (121)

HOWARD S. FRIEDMAN, *Headquarters, Aerospace Medical Division, Air Force Systems Command, Brooks Air Force Base, Texas* (223)

S. RAYMOND GAMBINO, *Englewood Hospital, Englewood, New Jersey, and Columbia-Presbyterian Medical Center, New York, New York* (55, 169)

ADRIAN HAINLINE, JR.,* *Duke University Medical Center, Durham, North Carolina* (143)

HANS HOCH, *Veterans Administration Center, Martinsburg, West Virginia* (159)

FRANK A. IBBOTT, *University of Colorado Medical Center, Denver, Colorado* (101)

ROY, B. JOHNSON, JR., *Scripps Clinic and Research Foundation, La Jolla, California* (159)

ALEX KAPLAN, *Biochemistry Department and University Hospital, University of Washington, Seattle, Washington* (245)

RODERICK P. MACDONALD, *Harper Hospital, Detroit, Michigan* (65, 237)

SAMUEL MEITES, *The Children's Hospital, Columbus, Ohio* (113)

* Present address: St. Luke's Hospital, Kansas City, Missouri.

ix

NATHAN RADIN, *The Rochester General Hospital, Rochester, New York* (91)

MIRIAM REINER, *District of Columbia General Hospital, Washington, D.C.* (257)

EUGENE W. RICE, *William H. Singer Memorial Research Laboratory, Allegheny General Hospital, Pittsburgh, Pennsylvania* (121, 231)

GUILFORD G. RUDOLPH, *Department of Biochemistry, Vanderbilt University, Nashville, Tennessee* (211)

LEONARD T. SKEGGS, JR., *Department of Medicine and Surgery, Veterans Administration Hospital, Cleveland, Ohio; and Department of Pathology, Western Reserve University, Cleveland, Ohio* (31)

AGNES STUMPFF, *St. Mary's Hospital, Reno, Nevada* (121)

RALPH E. THIERS, *Clinical Chemistry Laboratory, Duke University Medical Center, Durham, North Carolina* (131)

SEYMOUR WINSTEN, *Albert Einstein Medical Center, Philadelphia, Pennsylvania* (1)

BENNIE ZAK, *Wayne State University College of Medicine, Department of Pathology, Detroit, Michigan* (79)

PREFACE

Volume 5 of "Standard Methods of Clinical Chemistry" is an extension of the series initiated in 1953 by Dr. Miriam Reiner, and succeeded by three volumes under the guidance of Dr. David Seligson. The style of presentation is mainly theirs. The task of editing was simplified by their advice and encouragement and for this we express our gratitude.

The Editor and Editorial Committee of "Standard Methods," by *general agreement,* choose a method with respect to the appropriate criteria listed below:

1. The method has been published previously, therefore is not presented as new material.
2. The method may offer an unusually advantageous feature for determining the unknown, such as the Berthelot reaction for ammonia used in the analysis of urea nitrogen presented in this volume.
3. The method is "sound," has "stood the test of time," and is "widely used."
4. The method improves another one by overcoming objectionable features and by increasing desirable qualities such as sensitivity, precision, and reliability.
5. The method promotes the accuracy of analysis, e.g., by introducing a purer standard.

Ideally, if a method satisfies most of the pertinent criteria, it has also been investigated fully, many facets of knowledge pertaining to it have been examined, no secrets or hidden tricks are needed for its execution, and its shortcomings are understood.

As a result of the large number of chemists engaged internationally in the fruitful study of methods, it is hardly surprising that there are alternative approaches to the analysis of most substances of clinical importance. Consequently, this volume includes two ways for determining bilirubin and magnesium in serum, as well as two for total proteins in cerebrospinal fluid. The chapter on glucose is the fourth since the "Standard Method" series began. Phosphatase and cholesterol are each treated for the third time.

How does the clinical chemist know which procedure to select? When an alternative method is incorporated in "Standard Methods" because it is based on a different chemical or physical principle, the chemist must usually decide for himself whether the merit of this method justifies the abandonment of another method. Thus, he may determine chloride electronically by titration with silver ion—as presented in Volume 2—or by manual titration with mercury ion, in the presence of an indicator—as described in Volume 1. Both methods are "standard," and both have practical advantages. Evidently, for determining many substances there is not always a *best* method; the choice must be made arbitrarily.

On the other hand, some alternative methods are presented here because they are improved versions of older ones, or demonstrate a superior as well as a different analytical approach. Improvement and superiority may be the result of an increase in sensitivity, a more adequate instrumentation, a sounder chemical basis, a sharper specificity, or a wider applicability. If the advantages of one method over others are properly demonstrated in a chapter, the chemist should have little difficulty in making the correct choice.

The Editors have designated as *provisional* two methods in this text. Such designation implies reservations about a procedure preventing its complete acceptance as standard, but it has proved too valuable to be omitted. The two azobilirubin methods for bilirubin are included in this volume because they are superior to several others currently available. Nonetheless, they are labeled *provisional* because the methodology is undergoing close scientific scrutiny so that newer procedures may soon prove more acceptable.

With the publication of Volume 5, one hundred and two methods have thus far been presented in the series covering an impressive array of detailed subject matter in clinical chemistry. However, while dealing with *specific* topics, "Standard Methods" should not overlook the *general* aspect of clinical chemistry. A great deal of accumulated information cannot be appropriately linked to individual methods. The "Standard Methods" series would not be complete without a determined effort to record this information. The collection and preservation of biological specimens is a subject germane to clinical chemistry. Sources of error must constantly command the attention of the clinical chemist. Automatic analysis, by introducing versatile machines, is causing a mushrooming of revisions in methods. Volume 5 considers these subjects. Discussion of broad problems in clinical chemistry should be a continuing and enhancing feature of "Standard Methods."

The Editors are delighted to introduce a new role to this series: the presentation of specifications for pure standards. Volume 5 offers recommendations for preparing bilirubin and cholesterol. Let this be a milepost for the future!

Clinical chemists should be reminded periodically not to ignore the infant as a subject for study. The tendency of chemists to devise new procedures and modifications while disregarding the infant, causes an unnecessary multiplication of methods: micromethods must be added. A small child imposes two limitations: small blood volume, and difficulty accessible veins (often saved for fluid therapy). As a result, blood is obtained by "capillary" puncture. The volume available from this source, however, is limited; hence, when multiple procedures must be performed on a single specimen (see p. 113), ultramicroanalysis is necessary. The micromethod for glucose published in Volume 5 is, in good part, an effort to supplement the macromethod in the previous volume. The urea nitrogen technique, however, because it was introduced to clinical chemistry as a microprocedure, is suitable for patients of all ages.

The clinical chemist is sometimes called upon to perform determinations that are ideally in the province of the almost nonexistent clinical physicist. The Editors have included a chapter on "Osmolality of Serum and Urine" falling into this category. Misgiving arises because the chapter admittedly lacks a chemical basis, and discusses the principles rather than the details of a specific method—a chore left to manufacturers of osmometers. On the other hand, inclusion of this chapter broadens the scope of "Standard Methods" by furnishing the clinical chemist with authentic information on an unusual task which, in many instances, has become his responsibility.

We are indebted to The Children's Hospital of Columbus, Ohio, for offering the facilities, material, and moral support essential to the Editor's task.

Our thanks to the Submitters and Checkers in Volume 5! Only through their cooperation and dedication to clinical chemistry was the substance of "Standard Methods" made possible.

SAMUEL MEITES, *Editor-in-Chief*

Columbus, Ohio
February, 1965

CONTENTS

COLLECTION AND PRESERVATION OF SPECIMENS

Submitted by: Seymour Winsten, Albert Einstein Medical Center, Philadelphia, Pennsylvania

Introduction

Collection and preservation of specimens for tests are daily problems in the clinical chemistry laboratory. Blind decisions often must be made about the preservative or anticoagulant to be used as well as the relative stability of chemical constituents in the sample. Not only have many investigators disregarded these problems, but several have made diametrically opposite observations. This review does not resolve these problems, but presents a general consensus of accepted practice based on available information and laboratory experience.

The following are a few sound general rules for collection and preservation of biological specimens:

1. Except for metabolic studies, blood specimens should be obtained after an overnight fast, or at least 4 hours after a solid meal.
2. When the substance analyzed is in blood, serum is the preferred sample. Plasma or whole blood should be used only when specified.
3. Serum samples should be free of hemolysis, and should be separated from the clot within 2 hours of collection.
4. Assays should be performed within 5 hours after collecting the sample. If this cannot be done, then the sample should be refrigerated between 2° and 4°C. If the delay is greater than 24 hours, the sample is best preserved at a minimum of −12°C. Most substances, when frozen, are well preserved. Freezing and thawing denatures some proteins and lowers the concentration of other constituents. Thawing should be done rapidly, in a 37° to 45°C. water bath.
5. Urine, particularly for steroid assay, should be collected without preservatives and refrigerated during the time of collection.
6. Cerebrospinal fluid and other body fluids such as exudates and transudates should be analyzed promptly, or refrigerated at 4°C. Refrigerated specimens should be assayed without undue delay.

1

These rules stem from an idealistic concept of the operation of the hospital routine. Any clinical chemist can cite instances where the rules are inadvertently bent or broken. A major purpose of this article is to demonstrate where the bending of a rule will not result in poor laboratory practice.

Enzymes

The *in vitro* stability of these proteins has been extensively discussed (1). Enzyme activity may be lost at any temperature, though a great deal of this is prevented by rapidly assaying or freezing the specimen.

The following serum enzymes are stable at room temperature (20°–25°C.) for at least 8 hours, at 4°C. for at least 1 week, and frozen for 1 month or longer: glutamic-oxalacetic and glutamic-pyruvic transaminases, leucine aminopeptidase, lactic acid dehydrogenase, phosphohexoseisomerase, glutathione reductase, aldolase, isocitric dehydrogenase, adenosine deaminase, pseudocholinesterase, amylase, and lipase (1, 2, 3, 4, 5, 6, 7). Some of these may have a longer period of stability at room temperature, but it is generally poor practice to leave specimens overnight without refrigeration. Losses of enzyme activity in abnormal sera may be extremely variable, and may not follow any pattern.

Some enzymes are less stable. Ceruloplasmin oxidase loses activity at room temperature. No change, however, is observed within 2 days at 4°C. or within 2 weeks if frozen (8). Phosphoglucomutase is stable at room temperature for at least 8 hours, but the activity decreases after 2 days of refrigeration. Activity is not lost for 4 to 7 days when the specimen is frozen (9).

Alkaline and acid phosphatase differ in stability. Alkaline phosphatase remains stable at room temperature for at least 8 hours and for at least 1 week when frozen (10). A significant loss of activity may occur in frozen plasma after 6 months. Storage for 24 hours in a refrigerator may increase the activity by 5 to 10% (11). Acid phosphatase is very unstable at room temperature. Almost 50% of its activity may be lost within 5 hours at 25°C. Serum or plasma separated rapidly and frozen until analyzed does not lose appreciable activity (12). If this cannot be done, acidification of serum by the addition of 0.01 ml. of 20% (v/v) acetic acid per ml. of serum preserves acid phosphatase activity (1).

Either plasma or serum may be used for most enzyme determinations. The suitability of plasma depends upon the type of anticoagulant present. Anticoagulants containing enzyme inhibitors such as sodium fluoride, or metal chelates such as ethylenediaminetetraacetic acid (EDTA), must be avoided. Oxalate and citrate, but not heparin, inhibit amylase activity (13). Lactic acid dehydrogenase and acid phosphatase lose activity in oxalated plasma (1).

Hemolysis should be scrupulously avoided for most enzyme assays. Almost all serum enzymes are present in much higher concentrations in the erythrocytes than in plasma (1). Any visible amount of hemolysis will usually produce elevated levels. However, a small degree of hemolysis does not significantly alter phosphohexoseisomerase determinations (2).

For the assay of some enzymes, the need to use specimens from patients in the basal state has not been established beyond doubt. There is evidence, however, that certain enzymes are affected by diet and physical activity (14), while others deny this (15). Strenuous exercise may produce marked elevations in transaminase activity (16). Alkaline phosphatase should only be assayed on "fasting" specimens (17). Diurnal variation is noted with acid phosphatase activity (18). Though similar studies have not been performed with other enzymes, a good general rule is to collect specimens for these assays from patients who are as near the basal state as possible.

Erythrocyte levels of two enzymes, glucose-6-phosphate dehydrogenase and galactose-1-phosphate uridyl transferase, have been recently studied (19, 20). The latter enzyme should be collected in acid-citrate-dextrose solution if the test cannot be performed within 2 hours. Citrated blood may be refrigerated up to 4 days without appreciable loss of activity. A saline solution of heparin (1 mg./10 ml.) may be used as an anticoagulant if the test is done at once. Glucose-6-phosphate dehydrogenase should be assayed on fresh blood collected in heparin.

Elements and Electrolytes

Sodium, potassium, phosphorus, calcium, magnesium, iron, iodine, and chloride in serum are not affected by storage at room temperature for at least 8 hours, in a refrigerator overnight, or in the frozen state ($-12°C.$) for at least 1 year. If the sample is left at room temperature or even in a refrigerator for several days, precipitation of protein or growth of microbial contaminants may interfere with precise measurements (21). Furthermore, the serum should be rapidly separated from the clot to prevent exchange of electrolytes between cells and serum. This is especially critical for potassium, but if the separation must be delayed, the blood should be kept at room temperature. Goodman, Vincent, and Rosen (22) note a much greater increase in potassium levels in clotted specimens stored at 4°C. than at 25°C. These investigators believe this phenomenon results from the slowing down of the enzymatic processes at 4°C., so that they no longer carry potassium into the cell and cannot counteract the greater effect of its outward diffusion, due to the concentration gradient (22). Plasma collected in tubes containing large concentrations of heparin (200–300 units per 2–5 ml. of blood) can be left in contact with blood cells for more than 4 hours without any change in potassium concentration (23).

Gambino, in this volume (24), presents a detailed discussion on collecting and storing specimens for the evaluation of pH and P_{CO_2}. Similar techniques for handling specimens may be used for total carbon dioxide content.

Though plasma is satisfactory for some elemental or electrolyte analysis, serum is preferred. Obviously, plasma containing calcium-removing anticoagulants (oxalate, citrate, EDTA) cannot be used for calcium determinations, and plasma containing sodium or potassium salts of anticoagulants is unsuited for sodium or potassium determinations. Heparin at a concentration of 0.5 mg./10 ml. contains a small amount of sodium or calcium as the salt, and this may alter results. Heparin is the anticoagulant of choice for determining plasma carbon dioxide and for plasma and whole blood pH. If phosphorus is determined in plasma, the cells should be removed quickly: there may be an increase of inorganic phosphate owing to the hydrolysis of sugar phosphate esters in the erythrocytes and subsequent transfer of phosphorus to the plasma (1).

In 1959, Annino and Relman (25) reported that postprandial specimens from normal persons could be used for the evaluation of carbon dioxide content, chloride, sodium, potassium, and calcium. They also noted a minor, but significant, drop in the inorganic phosphorus level. Unfortunately, they did not extend their studies to hospitalized patients whose metabolism may be different from normal persons. Fasting specimens are not required for iodine determinations (26). Though diet does not affect serum iron levels, there is a large diurnal variation ranging from 15–100 µg./100 ml. Lower iron values usually occur in the late afternoon or early evening (27). Bowie, Tauxe, Sjoberg, and Yamaguchi (28), however, have recently shown that random diurnal variations occur in the serum iron levels of normal persons, and that there is no advantage in collecting samples at a set time each day.

In view of the high concentration of magnesium, iron, and potassium in the erythrocyte as compared to plasma, specimens showing even mild hemolysis cannot be used for the analysis of these constituents. Mather and Mackie (29) have recently devised a correction factor for overcoming the hemolysis problem in potassium determinations.

Hepatic Function Tests (Nonenzymatic)

There are many flocculation tests, but most institutions offer the cephalin-cholesterol flocculation, zinc sulfate, and thymol turbidity tests. Because these tests measure protein abnormality any collecting technique affecting serum protein may alter the results. Samples should be obtained from a fasting patient, as lipemia gives a falsely elevated result (1, 30). Though cephalin-cholesterol flocculation and thymol turbidity tests are best performed on fresh serum, overnight refrigeration of the serum does not change the results. Refrigerated serum to be used for the zinc sulfate turbidity test must remain in contact with the clot, otherwise the turbidity values are falsely low. Yonan and Reinhold (31) show this to be due to the loss of carbon dioxide from separated serum. The flocculation and turbidity tests are usually performed on serum. Caraway (1) indicates that oxalated plasma cannot be used for the thymol turbidity test. Frozen serum may be used for the cephalin-cholesterol test if the specimen is thawed rapidly (30).

The collection and storage of blood for conjugated and unconjugated bilirubin has been studied (1). If the serum specimen is protected from direct light, unconjugated (alcohol soluble) bilirubin is stable at room temperature for at least 8 hours. Exposure of the sample to sunlight, ultraviolet, or even the usual lighting in the laboratory may cause a 50% loss in unconjugated bilirubin content within 2 hours (32). Consequently, samples for bilirubin evaluation obtained from infants must be assayed quickly or protected from direct light.

Yonan and Reinhold (31) tested the effect of overnight storage at 5°C. in the dark on total serum bilirubin, and did not find a significant decrease in the level. Manufacturers of control sera state that indirect-reacting bilirubin is stable for at least 1 week in reconstituted lyophilized serum at 4°C. (33, 34). If an unmodified Malloy-Evelyn diazo method is used for determining bilirubin content, then any degree of hemolysis interferes in the azo-coupling reaction and results in decreased levels (1). In this volume, Gambino (35) describes a modified Jendrassik-Grof diazo technique that overcomes hemolysis. Spectrophotometric procedures may also be used to minimize the effect of hemolysis (36). Bilirubin may be determined in either serum or plasma.

The effect of diet on serum proteins is an unresolved problem. Annino and Relman (25) do not note a significant change in total protein content in normal persons, 45 minutes after a meal. The general consensus, however, is that the only reliable sample is the one obtained from a rested patient in a recumbent position. Whitehead, Prior, and Barrowcliff (37), find as much as 0.75 g./100 ml. increase in total protein in patients who are active and nonrecumbent, as compared to the same patients when at rest and recumbent. The increase is primarily in the albumin component.

Henry, Golub, and Sobel (38) state that no changes occur in serum stored at room temperature for 3 days, and that the stability can be extended to a month if the serum is refrigerated. Other investigators claim that an alteration in the α- and β-globulins occurs despite refrigeration (39). Freezing at $-12°C$. preserves serum proteins, but the samples must be rapidly thawed and thoroughly mixed before assay.

Hemolysis, lipemia, and bilirubinemia affect the analysis of total protein by the biuret technique, and the analysis of albumin by the biuret, the methyl orange, or the 2-(4'-hydroxyazobenzene)benzoic acid (HABA) method. Modification of existing procedures by adding a serum blank diluted with the appropriate buffer or reagent may control these interferences (40, 41, 42).

Most prothrombin time determinations are performed on oxalated plasma, though citrated plasma is also used (43). Schoen, Praphai, and Veiss (44) observed that prothrombin activity is stable at 22°–25°C. for at least 18 hours in oxalated plasma, as long as the collecting tube is not opened and has very little dead space. Small aliquots (0.1–0.2 ml.) of oxalated plasma rapidly lose prothrombin activity, even when refrigerated. Hemolysis has a negligible effect in the 1-stage prothrombin technique (1).

Bromsulphalein (BSP) dye retention assays should be made while the patient is fasting, since the capacity of the liver to eliminate this dye is altered in the postprandial patient. Specimens should be clear, nonicteric, and free of hemolysis. However, Seligson, Marino, and Dodson (45) present a technique for measuring BSP in serum in which these factors do not interfere, while Reinhold (46) and Gaebler (47) apply correction factors to overcome them. Oxalated or heparinized plasma, as well as serum, may be used for the BSP determination (1). A small decrease of BSP dye content (1.1–4.0%) occurs in specimens left at room temperature for 16 hours (48).

In this volume, Conn (49) reviews the essential factors in the collection and preservation of samples for performing blood ammonia tests.

Nonprotein Nitrogen

Among the most frequently performed tests in the clinical chemistry laboratory are those involving nonprotein nitrogen (NPN). Serum is the most suitable specimen for these determinations, though plasma and whole blood are also used. At very high concentration (above 150 mg./100 ml.) urea nitrogen diffusion from the erythrocyte decreases, resulting in lower than expected values for plasma (50). If a urease technique is used for determining urea nitrogen the sample must not be collected in a tube containing an enzyme inhibitor or anticoagulants containing ammonium salts. However, plasma or whole blood collected with these anticoagulants may be assayed for urea nitrogen by direct chemical methods such as the diacetyl monoxime and p-aminobenzaldehyde procedures (51, 52). The NPN content of the erythrocyte is approximately 1.76 times greater than that found in plasma (1). This is due to glutathione found almost exclusively in the red blood cell. As a result, NPN values obtained in whole blood are higher than in plasma or serum. Levels of uric acid in whole blood, on the other hand, are lower than in serum or plasma because uric acid in the red blood cells is approximately one-half that of the plasma (1). Nonprotein nitrogen constituents collected in a fluoride tube are stable for at least 2 days (53). Total NPN, urea nitrogen, creatinine, and uric acid are stable in blood for at least 1 week when refrigerated. In heparinized plasma there is a significant decrease in creatinine after 24 hours refrigeration (1). Levels of urea nitrogen, creatinine, and uric acid in plasma are not significantly decreased when stored at −10°C. for 6 months. The content of non-protein nitrogen, however, does significantly decrease in frozen plasma kept at −10°C. for 6 months (10).

Carbohydrate Metabolites

In most clinical laboratories glucose is the principal carbohydrate studied. However, recent interest has developed in lactic and pyruvic acid levels in blood because of their relation to vitamin B_1 deficiency (53).

Food intake affects blood glucose levels. Previous dietary history is significant in the interpretation of glucose tolerance curves. Due to metabolic effects, patients on high carbohydrate diets tend to have flat glucose tolerance curves, while subjects on a low carbohydrate diet may have abnormally elevated carbohydrate levels (54). It is necessary, therefore, to place a patient undergoing a glucose tolerance test on a balanced diet for several days prior to the test. Venous, capillary, or arterial blood glucose concentrations are nearly the same in fasting individuals. However, after the ingestion of glucose, concentrations in capillary blood are higher than those in venous blood (55). This is important in the interpretation of glucose tolerance tests or postprandial blood sugars obtained by capillary puncture.

At normal blood sugar levels the glucose content of plasma is about 15 mg./100 ml. higher than the glucose concentration in the red cells. This is primarily due to greater water content in plasma than in whole blood, resulting in the uneven distribution of glucose between the cells and plasma (1). If whole blood is used for an assay, an enzyme inhibitor, e.g., sodium fluoride (10 mg./100 ml.), should be included with the anticoagulant mixture to prevent an appreciable amount of glucose from being lost due to glycolysis at room temperature. The rapid preparation of a protein-free filtrate or the aqueous dilution of whole blood (1 + 80) may be used to control the effect of glycolysis on whole blood specimens collected without preservatives (1) (see p. 113). Fluoride preserves glucose in blood for 8 to 12 hours at room temperature, and for 48 hours in the refrigerator (53).

Serum or plasma lactic and pyruvic acids are erroneously elevated by hemolysis. These acids are relatively unstable as a result of metabolic processes in shed blood (53). If the analysis cannot be performed immediately the samples should be drawn without a tourniquet, the serum or plasma rapidly separated from the cells, then frozen until needed. The patient should be in a basal state since exercise causes a considerable increase in lactic acid (53).

Lipids

With the mounting interest in the relationship of lipids to atherosclerotic heart disease, the clinical chemistry laboratory must offer an increasing number and variety of determinations of blood lipids. These include free and total cholesterol, triglycerides, total free fatty acids (esterified and nonesterified), β-lipoproteins, and phospholipids. Either serum or plasma may be used since the usual anticoagulants do not affect lipid determinations. Unless metabolic studies are involved, the samples should be collected while the patient is fasting, since nutrients alter lipid levels.

Anderson and Keys (56) studied the stability of cholesterol in serum. Cholesterol is stable at 4°C. for only 48 hours in nonsterile serum, but is stable for 5 years at −20°C. Due to the variable effects of spontaneous hydrolysis and esterification in stored serum, specimens for cholesterol esters are stable only when frozen. Hemolyzed and icteric specimens should be avoided, though some investigators incorporate a reagent to adsorb the bilirubin (57).

Lipids, excepting cholesterol, change rapidly despite refrigeration. Therefore, unless the analyses are performed immediately the plasma or serum should be separated at once and frozen. Free fatty acids, for example, increase at room temperature because a small amount of lipoprotein lipase in normal plasma causes the slow evolution of fatty acids by hydrolysis of the esters (58). The administration of heparin, or the occurrence of a situational stress, e.g., by the mere drawing of the specimen, leads to artifactual elevation of free fatty acids (59). Blood should be drawn under standardized conditions and the patient mentally prepared. Hemolysis alters phospholipid levels as organic phosphate hydrolyzes rapidly to the inorganic form (60).

Urine Specimens

The proper evaluation of chemical constituents in a 24-hour urine specimen depends highly on good collecting technique. The following procedure provides adequate results:

Instruct the patient to discard the first voided specimen in the morning. Collect and combine in an appropriate vessel (3-liter size) all subsequent voidings for the next 24 hours. Carefully explain to the patient that any loss of specimen may appreciably distort the results of the analysis. Refrigerate the specimen until the entire 24-hour collection is sent to the laboratory.

No preservatives should be added to specimens used for bioassay of hormones such as chorionic and pituitary gonadotropins and estrogens. The specimen should be refrigerated and processed promptly upon arrival in the laboratory. Frozen aliquots preserve activity for at least 1 month. Urine for steroid analysis may be collected without preservatives as long as it is refrigerated. However, if the specimen cannot be refrigerated the addition of certain preservatives increases its stability. For example, 10 ml. of toluene or 10 to 15 ml. of concentrated hydrochloric acid can be placed in containers for urinary total 17-ketosteroids, pregnandiol, pregnantriol, hydrocortisone, and cortisone (61), whereas aldosterone is best preserved by adding 15 ml. of 6 N acetic acid and 5 ml. of 40% formaldehyde to the collecting vessel (62). If estrogens are to be evaluated chemically, 25 ml. of chloroform is a useful preservative (63).

Free catecholamines, norepinephrine, and epinephrine, as well as their metabolite, 3-methoxy-4-hydroxymandelic acid (VMA), are stable in acid urine for several days at room temperature. If possible, the samples should be collected with a sufficient quantity of concentrated hydrochloric acid so that the final pH is not greater than 3 (64). This may usually be accomplished by placing 15 ml. of concentrated hydrochloric acid in the collecting vessel. If the urine pH is higher than 3 when the specimen arrives in the laboratory it should be adjusted immediately with acid. Free catecholamines deteriorate at neutral or alkaline pH. The Gitlow (65) procedure for measuring VMA may give false elevations when alkaline urine is used. This is due to the development of interfering chromogens.

CREATININE AND CREATINE

Creatine and creatinine are very unstable even when refrigerated due to an equilibrium reaction between these substances in aqueous solution (66). This reaction is catalyzed from either creatinine or creatine toward an equilibrium state by hydroxyl ions, and toward creatine alone by hydrogen ions. The maintenance of a neutral pH, though quite difficult, is the best way to preserve these substances. Retardation of bacterial growth (usually leading to the production of ammonia and an alkaline pH) is accomplished by layering the urine with about 50 ml. of toluene and refrigerating the collection bottle. Once the sample is sent to the laboratory it should be analyzed immediately.

ELEMENTS

Sodium, potassium, and chloride are relatively stable in refrigerated urine. Samples are best analyzed soon after collection to avoid the precipitation of various constituents. When precipitation occurs, the specimen must either be filtered or the precipitate redissolved by warming or by mild acidification (21). When filtered, the residue must be analyzed.

Calcium and phosphorus are prevented from precipitation by adding about 40 ml. of glacial acetic acid to each collecting bottle (67). Urine used for copper and mercury determination may be collected in a borosilicate glass-stoppered bottle previously rinsed with nitric acid, or in a plastic bottle (67).

ENZYMES

Clinical interest has been recently directed toward levels of certain enzymes found in 8- to 24-hour urine specimens. Among these are lactic acid dehydrogenase, alkaline phosphatase, leucine aminopeptidase, and amylase (68 69, 70, 71). Alkaline phosphatase, amylase and leucine aminopeptidase may be collected without preservatives if the urine is refrigerated. Lactic acid dehydrogenase loses activity rapidly and should be analyzed immediately following urine collection. Urines for amylase, alkaline phosphatase and leucine aminopeptidase may be stored in a refrigerator for at least 48 hours, and in a freezer for 1 week.

MISCELLANEOUS ANALYSES

Porphyrins, Porphobilinogen, and Urobilinogen

These substances are best preserved at an alkaline pH (72, 73). This is accomplished by adding 5 g. sodium carbonate to the urine bottle. Oxidation of pigments or pigment precursors is prevented by layering the urine with 100 ml. of petroleum ether. Because light promotes oxidative deterioration of uro- and porphobilinogen, amber glass or plastic bottles should be used for the collection (73).

5-Hydroxyindoleacetic Acid (5-HIAA)

5-HIAA is best preserved in an acid medium. This is done by placing 25 ml. of glacial acetic acid and 25 ml. of toluene in the collecting bottle. These reagents prevent oxidation of 5-HIAA. Acetic acid should be used for the acidification because mineral acids may cause degradation of indole (74).

Reducing Sugars

Reducing sugars are best preserved by adding 4 or 5 ml. of toluene to the collecting vessel (75). Chloroform or formalin should not be used since they interfere with the detection of glucose (53).

Free Amino Acids

The 24-hour urine specimen for amino acids may be collected without preservatives (76). If the analysis is delayed longer than 2 hours, the entire specimen is preserved by saturating with thymol and storing at 4°C., or freezing without preservatives.

Closing Remarks

The preceding discussion cannot be considered an exhaustive study of the collection and preservation of biological specimens. Several topics are not covered, including vitamin assays, toxicological evaluations, blood hormonal tests, and collecting techniques for microchemistry. In addition, the ever increasing problem of the effects of drug therapy on chemical procedures merits a separate review.

There are certain areas where more information is particularly needed. Little attention has been paid to the preservation of the chemical constituents of body fluids other than blood and urine. As noted under the general rules, cerebrospinal fluid, transudates, and exudates should be studied promptly, or refrigerated at 4°C. However, apparently no one has actually investigated this matter within the last two decades.

Another area warranting further attention is the influence of storage and collection technique on *abnormal* specimens. A recent report, for example, states that lactic acid dehydrogenase (LDH) is better preserved in sera from hepatitis patients if stored at room temperature (20–25°C.) in contrast to 4°C., as a better storage condition for sera from normal subjects (77). This is apparently due to greater stability of one of the isoenzymes of LDH at room temperature rather than in a refrigerator. Similar studies of other enzymes, especially those separable into multimolecular fractions, are certainly desirable. The effect of frozen storage on various chemical constituents in urine should be more carefully investigated.

Since the proper collection and preservation of samples is as fundamental to good laboratory practice as the use of good analytical technique, a good way to conclude this review is by paraphrasing Hamlet's admonition to Horatio: "There is more to doing blood and urine assays, chemists, than there is in just your chemistry."

REFERENCES

1. Caraway, W. T., Chemical and diagnostic specificity of laboratory tests. *Am. J. Clin. Pathol.* 37, 445–464 (1962).
2. Bodansky, O., Serum phosphohexose isomerase in cancer. I. Method of determination and establishment of range of normal values. *Cancer* 7, 1191–1199 (1954).
3. West, M., Berger, C., Rony, H., and Zimmerman, H. J., Serum enzymes in disease. VI. Glutathione reductase in sera of normal subjects and of patients with various diseases. *J. Lab. Clin. Med.* 57, 946–954 (1961).
4. Fleisher, G. A., Aldolase. In "Standard Methods of Clinical Chemistry" (D. Seligson, ed.), Vol. 3, pp. 14–22. Academic Press, New York, 1961.
5. Wolfson, S. K., Jr., and Williams-Ashman, H. G., Isocitric and 6-phosphogluconic dehydrogenases in human blood serum. *Proc. Soc. Exptl. Biol. Med.* 96, 231–234 (1957).
6. Schwartz, M. K., and Bodansky, O., Serum adenosine deaminase activity in cancer. *Proc. Soc. Exptl. Biol. Med.* 101, 560–562 (1959).
7. Smith, R. L., Lowenthal, H., Lehmann, H., and Ryan, E., A simple colorimetric method for estimating serum pseudocholinesterase. *Clin. Chim. Acta* 4, 384–390 (1959).
8. Henry, R. J., Chiamori, N., Jacobs, S. L., and Segalove, M., Determination of ceruloplasmin oxidase in serum. *Proc. Soc. Exptl. Biol. Med.* 104, 620–624 (1960).
9. Bodansky, O., Phosphoglucomutase activity in human serum. *Cancer* 10, 859–864 (1957).
10. Walford, R. L., Sowa, M., and Daley, D., Stability of protein, enzyme, and nonprotein constituents of stored frozen plasma. *Am. J. Clin. Pathol.* 26, 376–380 (1956).
11. Bodansky, A., Phosphatase studies. II. Determination of serum phosphatase. Factors influencing the accuracy of the determination. *J. Biol. Chem.* 101, 93–104 (1933).

12. Davidson, M. M., Stability of acid phosphatase in frozen serum. *Am. J. Clin. Pathol.* **23**, 411 (1953).
13. McGeachin, R. L., Daugherty, H. K., Hargan, L. A., and Potter, B. A., The effect of blood anticoagulants on serum and plasma amylase activities. *Clin. Chim. Acta* **2**, 75–77 (1957).
14. Wang, C. C., and Appelhanz, I., A preliminary report on some extraneous factors that may influence serum glutamic oxalacetic transaminase level. *Clin. Chem.* **2**, 249–250 (1956).
15. Steinberg, D., Baldwin, D., and Ostrow, B. H., A clinical method for the assay of serum gultamic-oxalacetic transaminase. *J. Lab. Clin. Med.* **48**, 144–151 (1956).
16. Remmers, A. R., and Kaljot, V., Serum transaminase levels. Effect of strenuous and prolonged physical exercise on healthy young subjects. *J. Am. Med. Assoc.* **185**, 968–970 (1963).
17. Kaser, M. M., and Baker, J., Alkaline and acid phosphatase. *In* "Standard Methods of Clinical Chemistry" (D. Seligson, ed.), Vol. 2, pp. 123–131. Academic Press, New York, 1958.
18. Roubicek, M., and Winsten, S., Effect of routine rectal examination on the level of serum acid phosphatase. *J. Urol.* **88**, 288–291 (1962).
19. Anderson, E. P., Kalckar, H. M., Kurahashi, K., and Isselbacher, K. J., A specific enzymatic assay for the diagnosis of congenital galactosemia. I. The consumption test. *J. Lab. Clin. Med.* **50**, 469–477 (1957).
20. Beutler, E., Drug-induced hemolytic anemia. *In* "The Metabolic Basis of Inherited Disease" (J. B. Stanbury, J. B. Wyngaarden, and D. S. Frederickson, eds.), pp. 1031–1067. McGraw-Hill, New York, 1960.
21. Hald, P. M., and Mason, W. B., Sodium and potassium by flame photometry. *In* "Standard Methods of Clinical Chemistry" (D. Seligson, ed.), Vol. 2, pp. 165–185. Academic Press, New York, 1958.
22. Goodman, J. R., Vincent, J., and Rosen, I., Serum potassium changes in blood clots. *Am. J. Clin. Pathol.* **24**, 111–113 (1954).
23. Hultman, E., and Bergström, J., Plasma potassium determination. *Scand. J. Clin. Lab. Invest.*, **14**, Suppl. *64*, 87–93 (1962).
24. Gambino, S. R., pH and P_{CO_2}. This volume, p. 169.
25. Annino, J. S., and Relman, A. S., The effect of eating on some of the clinically important chemical constituents of the blood. *Am. J. Clin. Pathol.* **31**, 155–159 (1959).
26. Kingsley, G. R., and Schaffert, R. R., Protein-bound iodine in serum. *In* "Standard Methods of Clinical Chemistry" (D. Seligson, ed.), Vol. 2, pp. 147–164. Academic Press, New York, 1958.
27. Hamilton, L. D., Gubler, C. J., Cartwright, G. E., and Wintrobe, M. M., Diurnal variation in plasma iron level of man. *Proc. Soc. Exptl. Biol. Med.* **75**, 65–68 (1950).
28. Bowie, E. J. W., Tauxe, W. N., Sjoberg, W. E., Jr., and Yamaguchi, M. Y., Daily variation in the concentration of iron in serum. *Am. J. Clin. Pathol.* **40**, 491–494 (1963).
29. Mather, A., and Mackie, N. R., Effects of hemolysis on serum electrolyte values. *Clin. Chem.* **6**, 223–227 (1960).
30. Knowlton, M., Cephalin-cholesterol flocculation test. *In* "Standard Methods of

Clinical Chemistry" (D. Seligson, ed.), Vol. 2, pp. 12–21. Academic Press, New York, 1958.

31. Yonan, V. L., and Reinhold, J. G., Effects of delayed examination on the results of certain hepatic tests. *Clin. Chem.* 3, 685–690 (1957).

32. O'Hagan, J. E., Hamilton, T., Le Breton, E. G., and Shaw, A. E., Human serum bilirubin. An immediate method of determination and its application to the establishment of normal values. *Clin. Chem.* 3, 609–623 (1957).

33. Hyland Laboratories, 4501 Colorado Boulevard, Los Angeles 39, California.

34. General Diagnostics Division, Warner-Chilcott Division, Morris Plains, New Jersey.

35. Gambino, S. R., Bilirubin (modified Jendrassik and Grof). This volume, p. 55.

36. Meites, S., and Hogg, C. K., Direct spectrophotometry of total serum bilirubin in the newborn. *Clin. Chem.* 6, 421–428 (1960).

37. Whithead, T. P., Prior, A. P., and Barrowcliff, D. F., Effect of rest and activity on serum protein fractions. *Am. J. Clin. Pathol.* 24, 1265–1268 (1959).

38. Henry, R. J., Golub, O. J., and Sobel, C., Some of the variables involved in the fractionation of serum proteins by paper electrophoresis. *Clin. Chem.* 3, 49–64 (1957).

39. Laurell, C. B., Laurell, S., and Skoog, N., Buffer composition in paper electrophoresis. Considerations on its influence, with special reference to the interaction between small ions and proteins. *Clin. Chem.* 2, 99–111 (1956).

40. Reinhold, J. G., Total protein, albumin and globulin. *In* "Standard Methods of Clinical Chemistry" (M. Reiner, ed.), Vol. 1, pp. 88–97. Academic Press, New York, 1953.

41. Keyser, J. W., Rapid estimation of albumin and total protein in small amounts of blood serum. *Clin. Chim. Acta* 6, 445–447 (1961).

42. Chen, H. P., and Sharton, H., Evaluation of serum albumin by means of 2-(4'-hydroxy azobenzene) benzoic acid dye method. *Am. J. Clin. Pathol.* 40, 651–654 (1963).

43. Winsten, S., Evaluation of a photoelectric clot timer for the performance of prothrombin times. *Clin. Chem.* (In press.)

44. Schoen, I., Praphai, M., and Veiss, A., Storage stability and quality control of prothrombin time by means of the Quick method. *Am. J. Clin. Pathol.* 37, 374–380 (1962).

45. Seligson, D., Marino, J., and Dodson, E., Determination of sulfobromoph-thalein in serum. *Clin. Chem.* 3, 638–645 (1957).

46. Reinhold, J. G., Bromsulfalein tests. *In* "Medical and Public Health Laboratory Methods" (J. S. Simmons and C. J. Gentzkow, eds.), pp. 100–102. Lea & Febiger, Philadelphia, 1955.

47. Gaebler, O. H., Determination of bromsulfalein in normal, turbid, hemolyzed or icteric serums. *Am. J. Clin. Pathol.* 15, 452–455 (1945).

48. Winsten, S., Unpublished data.

49. Conn, H. O., Blood ammonia. This volume, p. 47.

50. Blackmore, D. J., Elder, W. J., and Bowden, C. H., Urea distribution in renal failure. *J. Clin. Pathol.* 16, 235–243 (1963).

16   SEYMOUR WINSTEN

2

51. Winsten, S., The use of a single manifold for sugar and urea determination on the AutoAnalyzer. *Ann. N.Y. Acad. Sci.* **102**, 127–136 (1962).
52. Michon, J., and Arnoud, R., Dosage colorimétrique rapide de l'urée sanguine. *Clin. Chim. Acta* **7**, 739 (1962).
53. Varley, H., "Practical Clinical Biochemistry," 3rd ed. Wiley (Interscience), New York, 1963.
54. Cantarow, A., and Trumper, M., "Clinical Biochemistry," 6th ed. Saunders, Philadelphia, Pennsylvania, 1962.
55. Reinhold, J. G., Glucose, *In* "Standard Methods of Clinical Chemistry" (M. Reiner, ed.), Vol. 1, pp. 65–70. Academic Press, New York, 1953.
56. Anderson, J. T., and Keys, A., Cholesterol in serum and lipoprotein fractions. Its measurement and stability. *Clin. Chem.* **2**, 145–159 (1956).
57. Babson, A. L., Shapiro, P. O., and Phillips, G. E., A new assay for cholesterol and cholesterol esters in serum which is not affected by bilirubin. *Clin. Chim. Acta* **7**, 800–804 (1962).
58. Gates, H. S., Jr., and Gordon, R. S., Jr., Demonstration of lipoprotein lipase in fasting human serum. *Federation Proc.* **17**, 437 (1958).
59. Cardon, P. V., Jr., and Gordon, R. S., Jr., Rapid increase of plasma unesterified fatty acids in man during fear. *J. Psychosomatic Res.* **4**, 5–9 (1959).
60. Sunderman, F. W., Comments on the estimation of phosphorus in serum. *In* "Lipids and Steroid Hormones in Clinical Medicine" (F. W. Sunderman and F. W. Sunderman Jr., eds.), pp. 25–27. Lippincott, Philadelphia, Pennsylvania, 1960.
61. Dorfman, R. I., and Shipley, R. A., "Androgens; Biochemistry, Physiology and Clinical Significance." Wiley, New York, 1956.
62. Sunderman, F. W., Determination of aldosterone in urine. *In* "Lipids and Steroid Hormones in Clinical Medicine" (F. W. Sunderman and F. W. Sunderman Jr., eds.), pp. 169–175. Lippincott, Philadelphia, Pennsylvania, 1960.
63. Green, J. W., Personal communication.
64. von Euler, U.S., "Noradrenaline." Thomas, Springfield, Illinois, 1956.
65. Gitlow, S. E., Ornstein, L., Mendlowitz, M., Khassis, S., and Kruk, E., A simplified urine test for pheochromocytoma. *Am. J. Med.* **28**, 921–926 (1960).
66. Edgar, G., and Shiver, H. E., The equilibrium between creatine and creatinine in aqueous solution. The effect of hydrogen ion. *J. Am. Chem. Soc.* **47**, 1179–1188 (1925).
67. Boutwell, J., "Clinical Chemistry. Laboratory Manual and Methods." Lea & Febiger, Philadelphia, Pennsylvania, 1961.
68. Wacker, W. E. C., and Dorfman, L. E., Urinary lactic dehydrogenase activity. I. Screening method for detection of cancer of kidneys and bladder. *J. Am. Med. Assoc.* **181**, 972–978 (1962).
69. Amador, E., Zimmerman, T. S., and Wacker, W. E. C., Urinary alkaline phosphatase activity. I. Elevated urinary LDH and alkaline phosphatase activities for the diagnosis of renal adenocarcinomas. *J. Am. Med. Assoc.* **185**, 769–775 (1963).
70. Goldbarg, J. A., and Rutenburg A. M., The colorimetric determination of leucine aminopeptidase in urine and serum of normal subjects and patients with cancer and other diseases. *Cancer* **11**, 283–291 (1958).

71. Budd, J. J., Jr., Walter, K. E., Harris, M. L., and Knight, W. A., Jr., Urine diastase in the evaluation of pancreatic disease. *Gastroenterology* **36**, 333–353 (1959).
72. Schwartz, S., Zieve, L., and Watson, C. J., An improved method for the determination of urinary coproporphyrin and evaluation of factors influencing the analysis. *J. Lab. Clin. Med.* **37**, 843–859 (1951).
73. Watson, C. J., Studies of urobilinogen. I. An improved method for the quantitative estimation of urobilinogen in urine and feces. *Am. J. Clin. Pathol.* **6**, 458–475 (1936).
74. Dalgliesh, C. E., The 5-hydroxyindoles. *Advan. Clin. Chem.* **1**, 193–235 (1958).
75. Sidbury, J. B., Jr., The nonglucose melliturias. *Advan. Clin. Chem.* **4**, 29–52 (1961).
76. Frame, E. G., Free amino acids in plasma and urine by the gasometric ninhydrin-carbon dioxide method. *In* "Standard Methods of Clinical Chemistry" (D. Seligson, ed.), Vol. 4, pp. 1–13. Academic Press, New York, 1963.
77. Kreutzer, H. H., and Fennis, W. H. S., Lactic dehydrogenase isoenzymes in blood serum after storage at different temperatures. *Clin. Chim. Acta* **9**, 64–69 (1964).

SOURCES OF ERROR IN CLINICAL CHEMISTRY

Submitted by: WENDELL T. CARAWAY, Flint Medical Laboratory, and Laboratories of McLaren General Hospital and St. Joseph Hospital, Flint, Michigan

Introduction

Accuracy is a prime objective in the clinical laboratory. Quantitative analyses are performed in the chemistry laboratory to a greater extent than in other clinical disciplines. Concurrent analysis of control specimens with each series of unknowns provides some assurance of *precision*. It must be assumed, however, that the initial series of observations on a control specimen were performed *accurately*. Should subsequent results exceed established limits of deviation from the mean, sources of error must be sought. Even a well established program of quality control will not insure accurate results on the individual specimen. Such factors as instability of components, presence of contaminants, preservatives, anticoagulants, interfering medications, hemolysis, lipemia, and hyperbilirubinemia may lead to grossly inaccurate results. It is imperative, therefore, that the laboratory director and the chemist-analyst be keenly aware of potential sources of error in chemical determinations.

The need for adequate facilities and qualified personnel does not require elaboration. Rather, an attempt has been made in this article to provide an outline of certain variables encountered in processing a sample from the initial collection to the final report. A brief checklist of possible interferences in the more common determinations of clinical chemistry is also presented. Supporting references for many statements may be found in a review by the author (1). Other articles on precision, accuracy, sources of error, and desirable criteria of analytical methods should also be consulted (2, 3, 4, 5, 6, 7).

Collection of Specimens

Blood for chemical analysis should be drawn preferably while the patient is in the postabsorptive state. Serum glucose, inorganic phosphate, thymol turbidity, and neutral fat are affected significantly by eating and moderate lipemia may interfere with other determinations.

19

Stress or exercise may result in increased excretion of catecholamines and 17-hydroxycorticoids, decreased urea clearance, a decrease in serum iron, and an occasional increase in serum cholesterol. Diet and nutritional status have an effect on glucose tolerance tests, serum lipids, and uric acid. Smoking, and drinking tea or coffee may be contraindicated in certain tests of gastric, kidney, or liver function. Changing from a recumbent to an upright position results in a temporary contraction in plasma volume with resultant increase in hemoglobin and plasma protein concentration.

Much more information is needed on the effect of various medications on chemical analyses. Anomalous results can sometimes be explained by reference to the patient's hospital record. If possible, the drug in question may be withheld for a few days and the test repeated.

Serum is preferred to plasma or whole blood for most chemical determinations. Virtually all components, including glucose, urea, and water are present in erythrocytes and plasma in different concentrations. In addition, the necessary anticoagulants in plasma may inhibit certain enzymes or interfere with determinations of calcium, sodium, potassium, and urea. Hemolysis must be avoided; substances present in erythrocytes in relatively high concentrations such as potassium, lactic dehydrogenase, acid phosphatase, and glutamic-oxaloacetic transaminase, invalidate results obtained on hemolytic serum when measuring the foregoing substances. Hemoglobin interferes directly with determinations of serum bilirubin, protein fractions, and lipase activity.

Collections of 24-hour urine specimens require the utmost cooperation of the patient, nursing staff, and laboratory. Inadequate preservative, loss of voided specimens, or inclusion of two morning specimens in a 24-hour period are common errors. Determination of total creatinine excretion can only be considered a rough guide to adequacy of collections. Five ml. of glacial acetic acid, added to the container at the start of a collection, is a satisfactory preservative for most determinations; bile pigments are more stable in alkaline solutions.

Processing

The composition of body fluids *in vivo* represents a state of dynamic equilibrium. Prompt analysis of specimens is desirable to avoid changes *in vitro*, e.g., hemolysis, glycolysis, production of ammonia, loss of carbon dioxide, changes in pH, inactivation of enzymes, hydrolysis of phosphate esters, shifts of electrolytes between erythrocytes and plasma, and diffusion of intracellular enzymes. If serum cannot be analyzed promptly, it should be separated from the clot and refrigerated or frozen.

Specimens must be well mixed before sampling. This applies especially to specimens that have been frozen and thawed, and to 24-hour urine collections. The pipet or other sampling device should be chosen so that its total capacity is used; thus, a 1 ml. sample should be delivered from a 1 ml. pipet and not from a 5 or 10 ml. graduated pipet. Subsequent reagents may often be added as aliquots from burets or graduated pipets without significant error.

A very common error often overlooked is failure to mix solutions adequately at various steps in an analysis. If the volume of solution in a test tube exceeds half the capacity, the contents cannot be mixed thoroughly by gentle tapping or swirling. Vigorous mixing, inversion, mechanical agitation, or use of a larger diameter tube are necessary. Construction of volumetric flasks is such that ten or more complete inversions and vigorous shaking are required for thorough mixing.

Specified time intervals in a procedure should be followed exactly unless the time effect has been shown to be unimportant. Clocks should be checked for accuracy. Methods described may be vague with respect to stability of final colors or to the critical nature of waiting periods at various steps in the procedure. Usually, these factors may be checked easily by the analyst.

Factory calibrations of pipets and other glassware are not infallible. If complete recalibration is not feasible, spot checks on new shipments are advisable. Intercomparison of pipets may be done by gravimetric, colorimetric, or radioisotope techniques; comparison of cuvets is accomplished by measuring the absorbance of a colored solution at the wavelength of maximum absorption.

Greater use of automation tends to standardize or eliminate most variables in processing. A major consideration in automatic analysis is to be certain that standards and unknowns are sufficiently similar in over-all composition so that the instrument will not discriminate during sampling, dialysis, or measurement.

Method of Analysis

Ideally, a method suitable for routine use should be specific, accurate, precise, rapid, inexpensive, employ stable reagents and standard apparatus, require small samples of blood, have low blank values, produce a definite endpoint or stable color, and require minimum working space.

Adequate recovery of pure standard added to a biological specimen provides some assurance that all the native compound is included in the measurement but satisfactory recovery studies alone do not imply that a method is specific for that substance. *Specificity* of a method is inferred by testing individually a large number of potentially interfering substances to determine the incidence of positive or negative effects. Certain drugs or their metabolic derivatives in serum and urine may inhibit enzymatic and chemical reactions. Such inhibition may be detected by performing analyses on mixtures of the specimen with a standard or known normal specimen, or by comparing results on diluted and undiluted specimens of the unknown. In the absence of an inhibitor, results should be directly proportional to the volume of specimen analyzed.

The most practical way to check the *accuracy* of a method is to analyze the same sample by other, independent, accepted methods. Inasmuch as an "accepted" method may itself be subject to question, entirely different chemical approaches should be used in comparing methods. Calcium, for example, may be determined independently by precipitation as the oxalate, phosphate, or chloranilate; titration with a chelating reagent; production of color with nuclear fast red or ammonium purpurate; emission spectrophotometry; and atomic absorption spectrophotometry. Bilirubin may be measured spectrophotometrically either directly or after conversion to biliverdin or a diazo derivative. All such comparisons should be performed repeatedly on samples containing low, normal, and high concentrations of the constituent under study. In addition to the multiple-procedure check to validate a method, samples may be analyzed in replicate by other persons; at different times; in separate laboratories; and with reagents and standards from new sources.

The contribution of the *blank* should be minimal. Methods based on the difference between a standard and a large or variable blank are inherently less accurate as they are dependent on measurement of a small difference between two large numbers. A suitable blank should approximate the composition of the unknown with the exception of the constituent being measured and should be treated in the same way as the unknown sample. Water may often be substituted for the sample in a blank determination after demonstration that all interference arises from the reagents. The effect of variations in the blank is minimized when the concentration of the unknown approximates that of the standard and is nil when these concentrations are equal. Care must be taken to insure that a component of the blank is not subtracted twice as is done, for example, in certain saccharogenic methods for the determination of amylase activity. In these methods two blanks are included to correct for reducing substances in serum and starch respectively. Since the copper and arsenomolybdic reagents alone contribute color to both blanks the net effect is to reduce the apparent amylase activity. This is corrected by combining the two blanks or by including a separate reagent blank.

Standards

With the exception of a few gravimetric or empirical procedures, results in clinical chemistry are compared ultimately to results on a primary standard of the pure constituent. Criteria of purity for biological compounds are not well established. Even with extensive purification many compounds readily undergo changes as a result of oxidation, hydrolysis, decomposition, absorption of water vapor, and other factors. Preparation of a pure organic standard is outside the scope of many laboratories but a few suggestions are appropriate. Purified material from a commercial source is further purified extensively by two or more independent procedures, one of which should include formation of a derivative. These preparations are then subjected to melting point determination and elemental analysis; determination of molar absorptivity in a specified solvent at a verified monochromatic wavelength, depth of solution and temperature; and determination of the ratio of absorbance at two wavelengths. Next, the material is subjected to one or more chemical reactions under exactly defined conditions and the absorbance so obtained is recorded as a secondary standard value. Finally, the purified material should be checked for stability under various conditions of storage both in solution and in the dry state. This general approach has been applied to establish criteria of purity for cholesterol (8) (see p. 91.)

When possible, comparisons should be made on standards prepared in pure solvent against those prepared in a biological medium, i.e., serum or urine. The latter preparation serves as an "internal" standard and will detect presence of interfering substances. If results on these two types of standards are different, it is advisable to prepare

the standards routinely in the biological material. Standards for sul-
fobromophthalein and Evans Blue dye, for example, should be prepared in serum since their spectral qualities are influenced by albumin. Even with this refinement, unpredictable pathological variations
in serum albumin affect the final results to some extent. Pooled serum
or urine used as solvent for standards should have an approximate
normal composition as determined by analysis or spectral absorbance
characteristics.

Various control serums available commercially are useful to permit comparison of one's own results with those obtained in other
laboratories. It is advisable to have material from more than one manufacturer available for comparison. The use of commercial control
serum as a primary standard is a compromise not consistent with good
chemical practice.

Reagents

Although reagent grade chemicals should be used whenever possible, the criteria for purity of reagents are not as stringent as for
primary standards inasmuch as minor impurities can be compensated
by blank determinations.. The major source of error encountered with
reagents is deterioration or change in properties with time. Factors
affecting the stability and characteristics of reagents include: inherent
instability, type of solvent, pH, light, temperature, oxidation, mold
growth, preservatives, evaporation, absorption of carbon dioxide or
water vapor, contamination with dust, contamination from corks or
bottle caps, dirty glassware, impurities in distilled water, leaching of
silicates from glass containers, binding of constituents to glass or polyethylene containers, and leaching of surface-bound constituents by
subsequent solutions.

Stability of reagents may sometimes be improved by change
of solvents, preparation of more concentrated standard solutions, and
by addition of a preservative. Refrigeration improves stability but care
must be exercised to avoid crystallization of components from solution; in addition, results obtained with chilled reagents may differ
from those obtained at room temperature with respect to enzyme assay, pH, or rate of color development.

Polyethylene is permeable to water vapor. Evaporation with increase
in concentration of reagents and standards occurs even in tightly-
stoppered bottles. As the volume of remaining liquid approaches 10%

or less of the total capacity, the ratio of surface area to volume increases concomitantly with the increased surface area of polyethylene available for diffusion, and evaporation becomes a significant factor. For this reason, small volumes of reagent should not be stored in oversize polyethylene bottles for extended periods of time. Increases up to 10% in concentration of a reagent have been observed under these conditions.

Polyethylene is not completely inert as evidenced by binding of dyes, stains, iodine, and picric acid. Binding of colorless reagents may go undetected only to manifest themselves by erratic behavior of subsequent solutions. Slow reduction of ceric and cupric ions has been observed in solutions stored in polyethylene bottles, and some solutions have developed significant fluorescence. On the other hand, alkaline solutions stored in glass containers gradually produce a soluble silicate that can interfere markedly in certain methods for determination of calcium.

Purity of distilled water is often taken for granted. Mechanical carry over of raw water in a still is usually detected by monitoring the conductivity of the distillate. Free chlorine should be tested for specifically as this strong oxidizing agent readily destroys bilirubin and uric acid. Trace elements, especially copper and fluoride, interfere with certain enzyme assays and other determinations. Water passed through ion exchange resin should be free of ions but may contain neutral organic matter.

While it is good practice to compare new reagents with the old to detect gross errors in preparation, it is never advisable to adjust the new solution to match an old solution that may have undergone evaporation, contamination, or deterioration. Absorption of carbon dioxide by alkaline solutions or buffers and absorption of water vapor by sulfuric acid lead to changes in composition. Consequently, new reagents should always be checked against primary standards or control serum.

Instruments

Increased use of instrumental methods of analysis requires some familiarity with sources of instrument error. Erratic behavior may occur from line voltage fluctuations; improper grounding; weak batteries, lamps, or phototubes; poor electrical connections, and so on. Most such difficulties are readily detected and repaired. Gradual changes in instrument performance are best monitored by a system of quality control with routine inclusion of standards and control specimens. A record of the performance of each instrument is a valuable aid in maintenance. This should include dates when new bulbs, tubes, batteries, etc., were installed; wavelength checks; readings on standard reference solutions; and other repair and maintenance data.

Some factors that affect the constancy of readings on spectrophotometers include: wavelength calibration; band width of light; improper focusing; aging light bulbs; stray light reflected from bulbs or lenses; and weak photocells. A concentrated solution of nickel sulfate, sealed in a cuvet and stored in the dark, is stable for years and is useful to check the stability of instrument readings. Maximum transmittance occurs at 510 mμ. A gradual decrease in transmittance with time suggests an increasing band width and loss of sensitivity. Other readings are taken at 400, 460, 550, and 700 mμ to monitor wavelength settings and status of stray light.

NOTE: Prepare the nickel sulfate solution by dissolving 40 g. of $NiSO_4 \cdot 6H_2O$ in freshly distilled water containing 1 ml. of conc. hydrochloric acid and diluting to 100 ml. Reserve as a blank a matched cuvet containing 1:100 dilution of conc. hydrochloric acid. Store the sealed cuvets in a cool dark place.

Because of the logarithmic nature of the absorbance scale on most spectrophotometers, the relative error in reading is minimal at 36.8% transmittance (T). An absolute error of 1% T at 10, 37, and 90% T corresponds to relative errors of 4.4, 2.8, and 10.8% respectively; readings should be taken between 20 and 60% T for minimum relative error. However, the absolute error in absorbance at 90% T is only one-ninth that at 10% T and this may justify the use of the higher portion of the scale in some cases.

The analytical balance is the ultimate basis for preparation of primary standards. It should be stationed in an area free of vibration, drafts, or sudden changes in temperature. The beam must be raised and lowered gently to avoid dulling of knife edges. Corrosive materials such as cyanide or iodine should not be weighed in open containers. A general rule is to strive for 0.1% accuracy in preparation of standards and 1% accuracy in preparation of reagent solutions. This philosophy would require that no less than 0.2 g. of standard should be weighed on a balance having a sensitivity of 0.2 mg.; conversely, it would not be necessary to weigh 10 g. of reagent with an accuracy greater than 0.1 g. A somewhat similar approach is used as to choice of container for preparation of solutions; i.e., final volumes of standard solutions should be accurate to 0.1% whereas 1% accuracy is sufficient for most reagents.

Environment

The environment in which a chemical reaction occurs includes the effect of temperature, pH, light, oxygen, type of solvent, ionic strength, and specific effects of salts, buffers, or organic compounds. These factors assume special importance in assays of enzyme activity and must be carefully specified in the method inasmuch as pure standards of enzymes are rarely available for comparison. Buffer capacity should be adequate so that the final pH of the reaction mixture is relatively unaffected by addition of serum or urine. It is desirable to record the pH of final reaction mixture at the temperature of incubation for enzyme assays. Temperature equilibrium occurs slowly in air as compared to a water bath; consequently, it is important to measure the temperature of the reaction mixture, rather than the temperature of the ambient air.

Control of pH and temperature are both important in measurements of turbidity reactions. Slight variations in temperature exert a marked effect on blood pH as compared to most buffer systems. Fixing of proteins on filter paper after electrophoresis should be done at a controlled temperature to avoid gross changes in dye-binding capacity.

Random and Determinate Errors

Random errors, also called "statistical" or "indeterminate" errors, vary nonreproducibly from one measurement of a set to the next. Slight variations in technique and apparatus throughout a given procedure result in different values which form a normal distribution about the mean. Assuming that only random errors exist, the mean value of all the measurements will closely approach the true value while the standard deviation will provide a concept of the precision of measurement.

Determinate errors, on the other hand, will tend to displace the mean value from the true value by either a constant or proportional amount. Contamination of a reagent with the substance to be measured, for example, would introduce a constant error of absolute magnitude. Solubility losses are often constant errors. Comparison of unknowns against a weak standard would introduce a positive proportional error whose absolute value is proportional to the amount of constituent being measured. Determinate errors can be evaluated by comparison of independent methods, by varying the amount of sample in a single method, and by preparation of new standards and reagents (3, 7). It is the aim of the analyst to eliminate determinate errors and to keep random errors to a minimum.

Mistakes

Gross errors can arise from mistakes. Incredible results, off by a factor of 10 or 100, have been encountered in reports. Common mistakes include: misreading or misunderstanding of directions; omitting a step from a procedure; using the wrong pipet, cuvet, factor, or wavelength setting; miscopying results; and making mistakes in arithmetic. The analyst should be aware of the importance of his work and pay close attention to details. He should recognize an improbable result and repeat the analysis on the original specimen. Heavy work loads, emotional stress, harassment, and poor working environment are not conducive to accurate work.

The ability to estimate mentally the result of a calculation improves with practice and should be encouraged to avoid the misplaced decimal point. Charts and nomograms are helpful to reduce calculations to a minimum. A permanent record of instrument readings should be maintained for reference and review by the laboratory supervisor. Explicit, precise, unambiguous, printed directions for each procedure, including notes on special precautions or problems apt to be encountered, are mandatory. To quote from Archibald (2): "If any ambiguity exists in the final directions some technician is almost sure to find it and proceed according to the unintended meaning."

Specific Sources of Error

The following checklist is a brief summary of common sources of error in analytical clinical chemistry. The list could be expanded almost indefinitely. Medications or physiological variations that cause true changes in concentration of a component have not been included.

Ammonium (blood): Ammonium in heparin; delay in analysis; production of ammonia by strong alkali action on blood.

Amylase: Contamination with saliva; inhibition by oxalate and citrate; poor buffering; variability in starches.

Barbiturate (UV): Salicylate interference.

Bilirubin: Weak standards; destruction by light and by free chlorine in water;

interference by hemoglobin (diazo); chromogens in methanol; dextran therapy (turbidity with methanol); carotenemia and lipemia (direct spectrophotometric).

Blood, occult: Uncooked meat, bromides and iodides in diet; tannic acid in guaiac preparations.

Bromide: Color in protein-free filtrates.

Bromosulfalein: Use of aqueous standards without protein present; interference by hemoglobin.

Calcium: Chelating agents in serum following therapy (EDTA); indicator errors; sequestering from glass surfaces; leaching from filter papers; incomplete precipitation or loss by washing precipitates.

Carbon dioxide: Exposure of sample to air; incorrect indicator and endpoint (titrimetric).

Carbon monoxide hemoglobin: Old or jaundiced blood (spectrophotometric).

Catecholamines: Urine not acidified immediately; poor recoveries; loss by oxidation; false positives with quinine, quinidine, and methyldopa.

Cephalin flocculation: False positives from unclean glassware; light; and inactivated serum.

Chloride: Bromide; indicator error; old serum; line voltage fluctuations (coulometric).

Cholesterol: Impure standard; poor control of time and temperature; effect of bilirubin; light sensitivity in blue spectrum (Liebermann-Burchardt); bromide; protein; weak sulfuric acid (ferric iron methods).

Glucose: Delayed processing of blood; nonspecific methods; interference by glutathione; interference by ascorbic acid (glucose oxidase); contamination with cotton or paper cellulose (anthrone type methods).

Hemoglobin: Insufficient cyanide (cyanmethemoglobin); traces of cupric ion (oxyhemoglobin); deterioration of standards; light sensitivity of reagent.

5-Hydroxyindoleacetic acid: False positives from acetanilide, methocarbamol, mephenesin; false negative from phenothiazine.

Icterus Index: Improper wavelength; carotenemia; lipemia; hemoglobin.

Iodine, protein-bound: Trace contamination with iodine or mercury; roentgenographic contrast media, bromosulfalein, or high concentration of inorganic iodide in serum.

Iron: Random contamination; incomplete recoveries; color in protein-free filtrates or extracts of serum; hemolysis.

17-Ketosteroids: Destruction during acid hydrolysis; incomplete extraction; color from urine extracts; high blanks; interference by meprobamate.

17-Hydroxycorticosteroids (Porter-Silber): Incomplete enzymatic hydrolysis; deteriorated phenylhydrazine reagent; high blanks; weak standard; delay in processing.

Lactic dehydrogenase: Hemolysis; inhibition by oxalate; loss of DPNH activity in reagent; nonlinear reaction methods; poorly reproducible standard curves; poor temperature control.

Lipase: Olive oil not emulsified; inactivation of enzyme; inadequate substrate (tributyrin; Tween); wrong indicator; inhibition by hemoglobin.

Lipids, total: Extraction of nonlipid material; loss by evaporation; incomplete evaporation of solvent; analytical balance (out of order; low sensitivity, inaccurate weights).

pH, Blood: Improper anticoagulant; exposure to air; delay in processing; insensitive pH meter; inaccurate buffer standards.

Phosphatase, acid: Delay in processing and analysis; inhibition by oxalate and fluoride; hemolysis; nonspecific substrate.

Phospholipid: Incomplete digestion; phosphoric acid in hydrogen peroxide.

Phosphorus, inorganic: Contamination from detergents; hemolysis; delay in processing; deterioration of sulfonic acid reagent (Fiske-SubbaRow); inhibition by mannitol.

Potassium: Variable concentration of sodium; hemolysis; tobacco smoke.

Protein: Methods based on different properties; lipemic serum; denaturation of albumin by salting-out methods; biuret reagent deteriorated; use of inadequate turbidity methods on cerebrospinal fluid; overheating of paper and variable staining by dyes (electrophoretic).

Sodium: Contamination from detergents or poorly cleaned glassware; unstable or low sensitivity instruments; unstable gas pressure.

Thymol turbidity: Wrong buffer and pH; poor temperature control; reagent deteriorated; lipemic serum; unstable reference standard.

Transaminase (GO): Hemolysis; nonlinear reaction methods; poorly reproducible standard curves; poor temperature control; high blanks.

T-3 Test: Elevation by Dicumarol; butazolidin; salicylate; diphenylhydantoin.

Urea (urease methods): Fluoride; loss of urease activity; inadequate buffering of filtrates; turbidity with Nessler reagent.

Uric acid: Increased values from ascorbic acid, methyldopa, or glutathione (colorimetric); inhibition of uricase by formaldehyde in standards; oxidation by free chlorine in water.

Urobilinogen: Oxidation by light; sodium acetate not saturated.

REFERENCES

1. Caraway, W. T., Chemical and diagnostic specificity of laboratory tests. *Am. J. Clin. Pathol.* **37**, 445–464 (1962).
2. Archibald, R. M., Criteria of analytical methods for clinical chemistry. *Anal. Chem.* **22**, 639–642 (1950).
3. Linnig, F. J., Mandel, J., and Peterson, J. M., A plan for studying the accuracy and precision of an analytical procedure. *Anal. Chem.* **26**, 1102–1110 (1954).
4. Ayres, G. H., Evaluation of accuracy in photometric analysis. *Anal. Chem.* **21**, 652–657 (1949).
5. Kingsley, G. R., The control and precision of clinical chemistry methods. *The Filter* **18**, (2), 14–18 (1956).
6. Lewin, S., Variations and errors in experimental investigations. *Lab. Pract.* **10**, 99–101, 162–164, 363–366, 408 (1961).
7. Blaedel, W. J., and Meloche, V. W., The theory of error and the treatment of quantitative data. *In* "Elementary Quantitative Analysis," Chapter 5, pp. 35–60, 550–582. Harper & Row, New York, 1957.
8. Radin, N., and Gramza, A. L., Standard of purity for cholesterol. *Clin. Chem.* **9**, 121–134 (1963).

PRINCIPLES OF AUTOMATIC CHEMICAL ANALYSIS

Submitted by: Leonard T. Skeggs, Jr., Department of Medicine and Surgery, Veterans Administration Hospital, Cleveland, Ohio; and Department of Pathology, Western Reserve University, Cleveland, Ohio

Introduction

A fully automatic, continuous method for performing colorimetric chemical analysis was described in 1957 (1). An instrument, the *AutoAnalyzer*, which employs the principles of this method was introduced later in the same year.[1] Since then, it has been successfully applied to a number of different determinations until today a large part of a laboratory's workload can be accomplished by this means.

The great majority of the hospital laboratories in this country and abroad have accepted the method, and are using it for one or more of their determinations. As a result of this experience, it has been found that it is at least as accurate, and certainly more reproducible than manual methods, as ordinarily executed. More important, the reliability is greater than that of manual methods. A technician performing a normal day's work by means of manual techniques is required to execute hundreds of operations. Each of these operations provides an additional possibility of error which may easily result in an incorrect analytical result.

The automatic recording of analytical results is valuable in providing written evidence as to the validity of the analysis and permits verification of calculations and results at any time.

One of the more important advantages which an automated laboratory enjoys is the ability to handle sudden and unexpectedly large workloads. Similar "heavy days" in laboratories using manual methods put an undue and fatiguing strain upon valuable technicians, which can lead to additional errors.

Nearly all laboratories which have adopted automatic methods have been able to increase their average workload and provide many additional services.

The amount of glassware required by automated laboratories is greatly reduced. This not only decreases the expense of the initial and

[1] Technicon Instruments Corporation, Chauncey, New York.

31

continuing purchase of glassware but, more important, reduces to a minimum the problem of glassware washing.

Finally, it is important to recognize that the trend toward rapidly increasing hospital laboratory costs may often be ameliorated, and the cost per test certainly reduced by the adoption of automatic analytical methods.

There is general agreement that all of those tests which are done in large numbers, such as urea nitrogen and glucose, should be automated. There is less agreement when the numbers of tests to be performed are small. It is probable that economic advantages disappear when the group falls to less than 5 or 10 per day. However, all of the other advantages remain. In addition, automation usually offers simpler, easier procedures which a suitably trained technician will prefer.

General Description

The techniques employed by the automatic method are fundamentally different from the conventional methods in that the entire process is continuous rather than batchwise. Samples and reagents are aspirated by means of a multiple proportioning pump, and are pro-

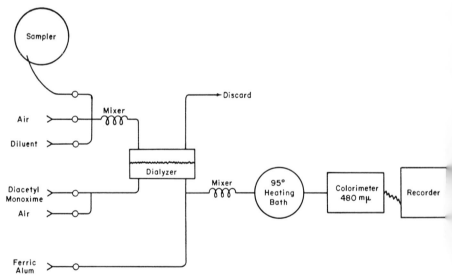

FIG. 1. Flow diagram of the method for determination of urea nitrogen. Small open circles represent individual tubes in the pump.

pelled through a flow circuit where they are mixed and processed by dialysis, heating, or other means, and a color produced in the stream which is in proportion to the concentration of the constituent being determined (Fig. 1). The stream is directed through a recording flowcell colorimeter where a continuous record of the analysis is produced. Unknown samples and standard solutions are introduced into the system for short intervals. The responses on the recorder are compared and the results of the analyses calculated.

Sample Integrity

It is necessary that samples following each other through the flowing system be prevented from intermixing in order that each might appear on the final record as a distinct individual, the magnitude of which is unaffected by the preceding or the following sample. This is largely accomplished by the introduction of air into the sample tube between successive samples. The air thus introduced forms a bubble which completely fills the lumen of the small tubing used in constructing the flow circuit and mechanically separates each sample as it is pumped into the flow circuit. In addition, the pressure of the air bubble against the inner wall of the tubing wipes the surface free of droplets of fluid which might contaminate the samples which follow.

It is advantageous to add additional bubbles of air to the sample stream when it is first diluted with reagents, and also to the stream of reagents which is to receive the diffusate in the dialyzer. It is thus provided that bubbles of air travel in the sample stream throughout the entire analytical train until they reach the colorimeter. By this means, the sample stream is actually divided into a larger number of individual segments or aliquots, the mechanical integrity of which is maintained by a bubble or air which precedes and another which follows. This procedure is of value in permitting a rapid change in the composition of the sample stream between successive samples (good wash), and is especially desirable when one of the components within the flow circuit does not respond instantly to a change in concentration.

Proportioning

Each sample segment is a separate aliquot and, in passing through the flow circuit, must receive and be thoroughly mixed with an accurately proportioned amount of each of the necessary reagents. Failure to accomplish this task properly results in disproportionate color development within the different segments, and in excursions of the recorder pen unrelated to sample concentration (noise). These difficulties may be avoided by the use of a proportioning pump with a positive and nonreciprocating flow (no suck-back), in conjunction with a correctly assembled flow circuit with properly functioning fittings which yield bubbles in a constant, regular pattern (bubble pattern), and produce sample segments of equal size.

Experience has shown that a bubble pattern, once established in a liquid stream, can be propelled through hundreds of feet of tubing for very long periods of time. Frequently, because of the physical character of the solution or the tubing material, it is desirable to reduce the surface tension and produce a wetting effect by the addition of a small amount of detergent. It should be noted that nitrogen may be used as the segmenting gas where chemical inertness is required. This expedient is required in the determination of amino acids with the ninhydrin reagent.

Mixing

Continuous mixing is accomplished, after the addition of reagents, by passing the stream through a tightly wound coil of tubing, usually about 2 cm. in diameter and lying on its horizontal axis. As the stream flows through the coil, the individual liquid segments are inverted many times as they pass repeatedly from the lower to the upper and the upper to the lower turns of the coil. Each inversion causes the portions of the solution which have a high specific gravity to fall through and partly mix with those portions having a lesser specific gravity. This process is repeated with each inversion until all parts have the same specific gravity and mixing is complete. The presence of air bubbles within the liquid stream is of positive value in this mixing process since it retains the sample aliquot and the proper proportion of reagents together within each liquid segment until they are thoroughly mixed.

It is frequently advantageous to mix rapidly and continuously a very small stream of a sample or reagent with a large reagent stream just prior to air segmentation. This can be accomplished by combining the two streams in a "Tee" connection and sending them through a short series of very small chambers joined with capillary tubing (Jet Mixer).

Occasional mixing problems which are too stubborn to be overcome by either of the foregoing methods have been solved by the use of a small "in-line" magnetic mixing flea.

Dialysis and Filtration

Serum or blood proteins interfere with the color development or other processes in a large number of clinical determinations and must be separated or removed. A continuous dialyzer has been successful in accomplishing this task. The sample stream, usually after dilution with reagents and segmentation with air, is passed on one side of a membrane while a second stream, also containing air bubbles, is passed in the same direction on the opposite side.

Although theoretically it is possible to remove all of a desired constituent from a sample stream by extended countercurrent dialysis, this is not generally practical. Co-current dialysis employing similar rates of flow on both sides of the membrane yields a more rapid washout. As a sample stream flows through the dialyzer, the diffusible constituents pass through the membrane and into the recipient stream. This stream, upon emerging from the dialyzer, contains the diffusible constituents of the serum in concentrations which, in every case yet examined, are in direct proportion to their concentrations in the sample stream.

Experience has shown that dialysis, when used in this fashion, is a simple and reliable analytical process. In only two instances have difficulties been encountered. It has been discovered that calcium dialyzes much more rapidly from serum than from the simple aqueous solutions usually used for standardization. The rates of dialysis are brought into agreement by conducting the dialysis in the presence of strong hydrochloric acid (2). A similar, although minor, problem has been discovered with respect to the chloride ion which dialyzes from diluted, acidified serum at a rate which is 3% less than from simple aqueous solutions. No practical difficulty has been encountered in the dialysis of sodium, potassium, or lithium, nor of any of the many nonelectrolytes which have been tested.

There are great differences in the rates of dialysis of molecules and ions which are of clinical interest. For example, sodium, potassium, and chloride dialyze exceedingly fast. Urea also dialyzes very rapidly, while the dialysis rate of creatinine is less, and glucose is relatively slow. Molecules larger than glucose dialyze at even slower rates. This is of great practical advantage in the determination of glucose and other substances as well since the dialysate contains very few interfering materials, and its relative "cleanness" increases the specificity of the method greatly. Certain molecules such as cholesterol will not pass through a cellophane membrane in an aqueous medium since they are virtually insoluble in water. Other compounds will not dialyze because they are tightly bound to the serum proteins.

The rate of dialysis of most substances increases with temperature. The effect appears to be smaller on those compounds which dialyze rapidly. In the case of those substances with slower rates of dialysis, the effect may be very pronounced. The rate of dialysis of glucose is known to increase greatly between 25° and 37°C. For this reason, a dialyzer which is being used for analytical purposes should be maintained at a constant, and preferably somewhat elevated, temperature.

The most practical membrane thus far discovered has been sheet cellophane which is easily obtained and very reliable. A much thinner grade has recently become available which permits faster rates of dialysis. Membranes of parchment, paper, collodion, and various plastics can and have been used, but are generally not advantageous.

A continuous filter has become available which can be used in place of the dialyzer. A stream of sample is continuously mixed with reagents capable of precipitating protein and is placed on a slowly moving strip of filter paper. Filtrate is continuously drawn through the paper with the aid of vacuum, and is mixed with a stream of reagents needed for further processing. The filter paper strip, after the filtrate has been withdrawn and while still wet with the remaining solution, is discarded.

Time and Temperature Control

Almost all clinical determinations require a period of time during which a chromogenic, enzymatic, or other type of reaction is allowed to occur. These are easily provided for by passing the solution through a coil of tubing of sufficient length to effect the desired time delay. If a certain temperature is required for the reaction to occur, this may be obtained by immersion of the coil in a bath which is thermostated at the desired temperature.

As a stream of liquid segmented by air bubbles is passed through a coil immersed in a bath maintained at an elevated temperature, the bubbles increase in size. This expansion will continue until the partial pressure of the air plus the vapor pressure of the liquid equals the barometric pressure. Thus it is obvious that baths must be operated at temperatures somewhat less than the boiling point of the liquid being used or explosive expulsion of the coil contents will occur. Moreover, the temperature should be sufficiently low so that unnecessarily great expansion of the bubbles is prevented, which would increase the rate of passage of the liquid through the coil, and thus reduce the heating time. For this reason, baths used to obtain temperatures near the boiling point of water are usually operated at about 95°C. since many reagents are dilute aqueous solutions. In some cases, the concentrations of solutes such as sodium chloride or sulfuric acid are high, thus reducing the vapor pressure and minimizing the effect of heating on bubble size.

Colorimetric reactions, as performed by conventional methods, are nearly always carried to completion. This is also desirable when the reactions are performed automatically in order to attain the highest sensitivity. Since the transit time through a reaction coil is automatically and accurately controlled by the rate of flow of the stream, it is entirely practical to stop the reaction prior to completion. In some cases, the total analysis time can be very appreciably shortened by this expedient.

Colorimetry

After color development, the air bubbles are removed from the stream by either decantation or by controlled aspiration from the upper arm of a "Tee" fitting. In this manner, the liquid segments, which up to this point have represented individual aliquots of the original sample, are combined, thus forming a solid stream of liquid. Small differences in color which may have existed between the segments are thus minimized by this automatic integration process. The stream is passed without delay through the colorimeter cell where the optical density of the stream is continuously measured and the results recorded.

It is desirable that the colorimeter cell be small and possess a shape which will allow a rapid washout in order that the colorimeter-recorder system will respond rapidly to a change of color of the sample stream. At the same time, it is also desirable that the path of the light beam through the cell be as long as possible in order that the maximum electrical signal be obtained, or sensitivity achieved for a

given change in color. These objectives seem to be best obtained by
a tubular-shaped cell with inlet and outlet connections for the flowing
stream at either end, and with the light beam arranged to pass through
in a longitudinal direction.

Other Endpoint Measurements

Any analytical "endpoint" device which can be adapted to continuous measuring and recording can be incorporated within the
automatic analysis system. A fluorometer, for example, has been used
for the determination of magnesium (3), and a flame photometer
for the determination of sodium and potassium (4).

The use of automatic equipment for flame photometry offers certain
inherent advantages. The sample may be injected directly into the
flame unit by the proportioning pump, thus permitting most efficient
and controlled use of the sample. At the same time, a protein-free
dialysate is used which greatly reduces the flame background emission
and also eliminates clogging of the capillary used for introduction of
the sample into the flame. Moreover, lithium is added to the serum
diluent, thus providing automatic internal standardization of both the
dialysis and the flame photometric processes.

Other Processes

Several other analytical processes have been adapted to continuous
operation and used automatically. These include digestion, distillation,
and extraction. Although these methods have received considerable
use in industrial or research applications, they have not received significant clinical use.

General Precautions

In theory, any manual colorimetric procedure can be adapted to the
automatic method. In actual practice, the first attempt to automate
any manual method is nearly always successful. However, it is usually
necessary to adjust the flow rates, reagent concentrations, heating
temperatures, and reaction times, in order to achieve linearity, the
desired sensitivity, low noise level, and best possible washout. For-
tunately, this has been done for many methods so that there is at
present one or more methods available for each of the common clinical
determinations.

However, this does not absolve the clinical chemist of his responsibility to check the existing methods, to improve them if necessary, or to develop different ones if it appears to be desirable.

Those persons who are in charge of clinical laboratories using AutoAnalyzers should periodically verify the condition of the pump, the age of the dialyzing membrane, the freshness of the reagents, standardization of the method, the character of the calibration curve, and the reproducibility of the method, together with its ability to yield the label values of commercial control serums. He should also take care that the noise level under steady state conditions is not too high, and that washout is as rapid as can be expected. Sample records illustrating good function should be posted near the equipment in order to invite comparison.

Sampling Methods

A common fault found in newly described automatic methods and often in the application of established methods is the use of too high a rate of analysis. The rate should be consistent with the quality of washout. It should be remembered that a sample which is aspirated continuously will yield a steady state signal at the recorder (Fig. 2). For routine analytical purposes, samples need only be pumped for a

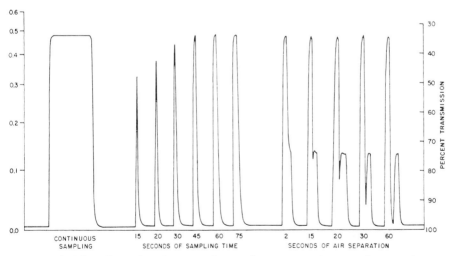

Fig. 2. The effect of various sampling and separation times on the recorder response.

period long enough to yield momentarily a steady state value at the recorder. Care should be taken that the sampling time is not reduced below this point or the response will be a spike, the height of which is unduly dependent upon the sampling time and quality of the washout as well as other factors unrelated to the concentration of the desired constituent.

It is easily possible to allow sufficient time between samples so that no sample will be affected by the one which precedes it (Fig. 2). If, in the interests of achieving a higher rate of analysis, this consideration is overlooked, then one must be prepared to observe all records carefully for samples which may have been affected by those which preceded it, and to analyze these again if necessary. Finally, records should be carefully examined for the presence of "spiking" responses or peaks which indicate short aspiration time due to an insufficient amount of sample, clots, or other sampling problems.

It is occasionally desirable to separate samples with a single bubble of air, a portion of water, and a second bubble of air. This technique may be automatically accomplished with a recently developed sampling unit. The procedure provides additional cleaning for the short piece of tubing connecting the pump and the sampler, and may be worthwhile when samples having very great differences in concentration are encountered. In addition, the water wash provides a constant flow of liquid during those periods when a sample is not being aspirated. This may be important when the stream must be propelled through an extended flow circuit and a constant flow rate is essential. Finally, the procedure insures that a constant dilution of reagents is maintained at all times and that a valid reagent blank is recorded. Whether such a precaution is necessary may be easily determined by comparing the baselines obtained during the sampling of air and of water.

Calculations

After the record of analyses is obtained, the calculations must be correctly performed and the results accurately transcribed. The calculations may be done directly on an overlaid template or "chart reader," or by means of semilogarithmic paper. In either case, the work must be carefully performed in order to prevent errors. It is possible to obtain equipment which will permit automatic calcula-

tion and digital printout of final results. It must be remembered, however, that the printout does not eliminate the need for the recorder
which indicates the function of the analyzer and verifies the accuracy
and reliability of the printed result.

Multiple Analytical Systems

There is considerable advantage in combining two or more methods
within a single flow circuit. For example, urea nitrogen and glucose

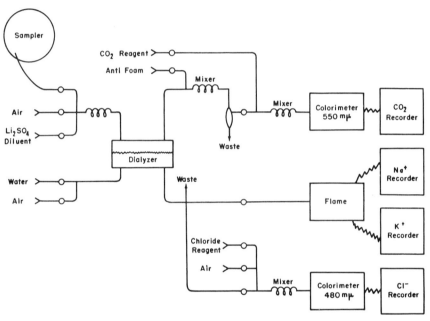

FIG. 3. Flow diagram of the method for the determination of sodium, potassium,
chloride, and carbon dioxide.

are very often combined and run as one method yielding two results
from a single sample. Creatinine and urea nitrogen have also been
combined, as well as calcium and phosphorus.

A most useful combination has been the four electrolytes sodium,
potassium, chloride, and carbon dioxide (Fig. 3). In order to effect
this combination, one aliquot stream of dialysate is pumped into the
burner of the flame photometer, thus permitting the simultaneous
recording of sodium and potassium. A second aliquot stream is used

for the determination of chloride. Carbon dioxide is determined by acid liberation of the gas from the nondialyzable stream, followed by reabsorption in a specific color reagent.

The flow circuit may be extended further. Urea nitrogen and glucose may also be determined on additional aliquots of the dialysate which are usually sent to waste. Total protein may be determined on an aliquot of the liquid phase of the nondialyzable stream after it emerges from the dialyzer. A separate small sample can be drawn directly from the sample cup for the colorimetric determinations of albumin. Thus a total of eight determinations may be performed on a single sample. Only one sample plate need be loaded and one identifying list prepared in order to accomplish this task which constitutes at least 60% of the workload of the usual hospital laboratory. The expense involved in performing the additional unrequested work according to this system is less than that of preparing separate aliquots for individual determinations according to the requisition slip. The extra analytical data does no harm, and frequently may be of very great value to the physician and his patient.

Systems are now under development which will perform all eight determinations and record the results from each sample on an individual piece of paper in calculated form suitable for immediate transmission to the requesting physician (5).

It is believed that multiple systems of analysis of this sort will be improved and extended to many other determinations. This should be of value not only in the laboratory diagnosis and treatment of patients, but also in the screening of the general population, and in the establishment of individual norms.

REFERENCES

1. Skeggs, L. T., Jr., An automatic method for colorimetric analysis. Am. J. Clin. Pathol. 28, 311–322 (1957).
2. Kessler, G., and Wolfman, M., An automated procedure for the simultaneous determination of calcium and phosphorus in serum and urine. Clin. Chem. 8, 429 (1962).
3. Hill, J. B., An automated fluorometric method for the determination of serum magnesium. Ann. N.Y. Acad. Sci. 102(1), 108–117 (1962).
4. Isreeli, J. Pelavin, M. and Kessler, G., Continuous automatic integrated flame photometry. Ann. N.Y. Acad. Sci. 87(2), 636–649 (1960).
5. Skeggs, L. T., Jr. and Hochstrasser, H., Multiple automatic sequential analysis. Clin. Chem. 10, 918–936 (1964).

BLOOD AMMONIA*

Submitted by: HAROLD O. CONN, Veterans Administration Hospital, West Haven, Connecticut, and the Department of Internal Medicine, Yale University School of Medicine, New Haven, Connecticut

Checked by: GORDON D. DUNCAN, Charlotte Memorial Hospital, Charlotte, North Carolina

DAVID SELIGSON, Yale University School of Medicine, New Haven, Connecticut

M. KENDALL YOUNG, JR., Woman's Hospital, Detroit, Michigan

Introduction

The blood ammonia[1] concentration has been widely used in the diagnosis and management of patients with liver disease. Methods for the measurement of the blood ammonia concentration have been based on the alkaline conversion of ammonium ion to ammonia gas, its isolation in an acid medium by aeration (2), distillation (3) or diffusion (4), and its quantitation by colorimetric or titrimetric techniques.

In 1935 Conway introduced a simple, reliable method in which diffusion and microtitration were carried out in a single porcelain diffusion dish (4). This technique, or modifications of it, is widely used. Unfortunately, these modifications in technique have resulted in an extremely broad range of reported normal blood ammonia values (5, 6). Seligson and Hirahara introduced a method (1) similar to Conway's in principle, but simpler and more adaptable to clinical use. Commonly used modifications of this technique have also resulted in discrepancies in the reported range of normal blood ammonia levels (1, 7, 8).[2]

* Based on the method of Seligson and Hirahara (1).

[1] Although ammonia exists almost entirely as ammonium ion at the pH of blood, the term blood ammonia is generally used to represent the total ammonia-ammonium content of blood.

[2] It should be emphasized that two "Seligson" methods are currently used for the determination of blood ammonia. The first, which was described by Seligson and Seligson in 1951 (9), is a general microdiffusion technique designed to measure ammonia in urine and other fluids low in protein content. This method uses K_2CO_3 for alkalinization. The second method (1) is designed specifically to measure ammonia in blood and other biological fluids with high protein content. Alkalinization is carried out with a mixture of K_2CO_3 and $KHCO_3$ which maintains a lower pH. With this buffered alkalinizing powder ammonia diffuses

43

Principle

Ammonium ions in blood are converted to ammonia gas by alkalinization with a buffered mixture of potassium carbonate and potassium bicarbonate. The reaction is carried out in a diffusion bottle which is sealed by a rubber stopper through which a glass receiving rod projects. The ammonia gas formed during a timed period of horizontal rotation of the bottle diffuses into a small volume of sulfuric acid distributed as a thin film over the tip of the receiving rod. The ammonia absorbed by the acid is transferred from the receiving rod directly to Nessler's solution and determined colorimetrically.

Apparatus

1. Diffusion bottles, 50 ml. capacity, 43 × 73 mm. with 13 mm. inlets.[3]

2. Rubber stoppers for diffusion bottles.[4]

3. Glass receiving rods, 6 mm. diameter, 80 mm. length with a constriction 12 mm. from one end. The short end of each rod is ground so that when it is dipped into acid it retains a thin film over its surface. The glass rod is mounted in a hole bored through the center of the stopper. When placed in the inlet of the diffusion bottle the receiving rod projects half-way into the bottle, and at the same time seals it (see Fig. 1).[5]

4. Stainless steel mixing rods, approximately 8 mm. in diameter and 40 mm. in length. These rods blend the reaction mixture and increase the diffusion surface.[6]

---

from blood at the same rate as from aqueous standards. The diffusion rates from blood and aqueous solutions differ when unbuffered K_2CO_3 is used. Some investigators have used the original Seligson-Seligson method to measure ammonia in blood. Unfortunately, it is not satisfactory for this purpose, and should be used only for the measurement of ammonia in urine.

[3] No. S-18B. T. C. Wheaton, Co., Millville, New Jersey. Scientific Industries, which advertises equipment for a modification of the Seligson–Seligson ammonia method, provides bottles and other accessories of different size which may alter diffusion rates and other variables.

[4] Corkage B, S-46 plug, T. C. Wheaton Co., Millville, New Jersey. One hole size 00 Neoprene stoppers may also be used for this purpose. However, some types of natural rubber stoppers cause opalescence of Nessler's solution.

[5] Macalaster Bicknell of Conn., 181 Henry St., New Haven, Connecticut.

[6] T. C. Wheaton Co., Millville, New Jersey.

FIG. 1. Seligson–Hirahara microdiffusion system. The horizontal diffusion bottle on the left contains 2 steel mixing rods and alkali powder. The receiving rod is in position. The vertical cuvet contains 4 ml. of Nessler's solution in which the receiving rod is immersed.

5. *Rotator.* This device is a vertically mounted wheel on which are placed 18 spring clips to hold the diffusion bottles in a horizontal position 6 to 12 cm. from the center of the wheel. The wheel is rotated at 50 rpm by a small electric motor.[7]

6. *Porcelain scoop* to deliver approximately 3 g. of alkali mixture.

7. *Automatic pipet* for rapid and accurate filling of cuvets with Nessler's solution (optional).[8]

Reagents

1. *Alkali mixture.* Mix thoroughly 2 parts by weight of potassium carbonate ($K_2CO_3 \cdot 1\frac{1}{2}H_2O$) and 1 part $KHCO_3$. An electric blender may be used for mixing. Do not powder the mixture. Store the mixture in sealed containers as it tends to clump in humid weather. It is convenient to place 3-g. aliquots of the alkali mixture directly into rubber-stoppered diffusion bottles for storage until used.

2. *1 N sulfuric acid.* Mix 1 vol. concentrated H_2SO_4 and 35 vol. deionized water.

3. *Nessler's reagent of Vanselow.* Dissolve 45.5 g. HgI_2 and 34.9 g. KI in as little water as possible (approximately 15 ml.). Add 112 g. KOH in 140 ml. water and dilute to 1 l. Let stand 3 days. Dilute 5:100 with deionized water before use. Commercial preparations may vary from batch to batch and sometimes contain impurities which cause the appearance of a fine white precipitate on the addition of ammonium sulfate. The vapor of acetone and other organic solvents may also cause precipitation in Nessler's solution; hence, such substances should be removed from the work area.

4. *Ammonium sulfate, stock standard, 0.300 mg. NH_3-N/ml.* Weigh 1.415 g. of $(NH_4)_2SO_4$ and dilute to 1 l.

5. *Ammonium sulfate, working standard, 3μg. NH_3-N/ml.* Prepare on the day of use by diluting the stock standard 1:100.

6. *Heparin solution.*[9] Although the ammonia concentration of commercial brands of heparin varies widely (10), the increase in blood ammonia concentration caused by the use of dilute heparin solution (10 mg./ml.) as anticoagulant is insignificant. Concentrated heparin

---

[7] Jewett Products, Racebrook Rd., Woodbridge 15, Connecticut, or T. C. Wheaton Co., Millville, New Jersey.

[8] "Tip-A-Tip" automatic pipet. Macalaster Bicknell.

[9] Sodium heparin solution, Liquaemin® (Organon, Inc., West Orange, New Jersey), 10 mg/ml. has proved consistently satisfactory.

solutions, however, may contain inordinately large amounts of ammonia which may increase blood ammonia levels artifactitiously (10).

NOTE: Anticoagulation with disodium ethylenediaminetetraacetate (1 to 2 mg./ml. of blood) is also satisfactory.[10] Potassium oxalate and sodium fluoride, which induce falsely high blood ammonia concentrations, should not be used for anticoagulation (11, 12).

NOTE: All reagents should be of analytic grade. Deionized water should be used in making all solutions.

Obtaining and Storing Specimens

Prepare syringes for drawing blood by wetting the barrel with heparin solution and expressing the excess. Arterial or venous blood may be used. Draw venous blood with a minimum of stasis. Do not have the subject clench his fist repeatedly because exercise may increase the blood ammonia concentration (13). Blood should be kept in the syringe until analyzed.

Analyze the blood within one-half hour of the time of collection. After this period there may be an increase of variable magnitude in blood ammonia concentration (14, 15). The stable period may be increased to an hour or more by refrigeration.

NOTE: Studies in the Submitter's laboratory have shown that if fresh blood after withdrawal is frozen immediately in liquid nitrogen or a mixture of ethanol and dry ice, the ammonia concentration will remain constant for 3 days, after which a slow increase occurs. The blood, however, must be stored in the frozen state ($-15°C.$), thawed rapidly at $38°C.$, and analyzed immediately after thawing (16). This frozen storage technique permits the preservation of blood specimens drawn at night or on weekends, or the transportation of blood specimens to central laboratories from hospitals at which blood ammonia determinations are not available.

Procedure

Prepare the diffusion bottles in advance by adding 3 g. of alkali mixture and 2 steel mixing rods to each. Before drawing blood, place the diffusion bottles in a horizontal position in the order of analysis

[10] Vacutainer #3204Q, Becton-Dickinson, Inc. (approximately 7 mg. of powdered disodium EDTA) contains a negligible amount of ammonia. The tripotassium ethylenediaminetetraacetate solution (Vacutainer #3204 QS, Becton-Dickinson, Inc.), however, should not be used because different batches have contained variable amounts of ammonia which falsely increased blood ammonia levels by as much as 90 μg./100 ml.

and rotate gently by hand to break up clumps of alkali powder and to align the steel rods side by side.

Perform all analyses in duplicate. Pipet 2.0 ml. of deionized H_2O into each of 2 diffusion bottles to serve as reagent blanks. Similarly pipet duplicate 2.0 ml. aliquots of blood and working standard. Take care to avoid wetting the neck of the bottles. Cap each bottle with a solid rubber stopper immediately after pipetting and place it on the rotator according to prearranged order. When pipetting has been completed rotate the bottles for 1 minute to facilitate mixing of the blood and alkali powder. Remove the bottles from the rotator in the order of analysis, replace the rubber stoppers with glass receiving rods which have been dipped into sulfuric acid up to the constriction, and return the bottles to the rotator.

NOTE: After the solutions have been pipetted, the bottles must remain horizontal at all times. The rubber stoppers can be firmly placed and easily removed with a slight rotating movement.

Rotate the bottles at 50 r.p.m. for 20 minutes. During the diffusion period pipet 4.00 ml. of Nessler's solution into appropriately labeled cuvets for the unknowns, standards, and blanks. At the end of the diffusion period carefully remove the receiving rods from the bottles (avoid touching the sides!) and immerse them directly into the Nessler's solution (see Fig. 1). The rubber stoppers on the rods permit mixing by inversion. Color development is complete in 5 minutes. The color is stable for several hours thereafter. Read the absorbances of the solutions against a water blank in a photoelectric colorimeter or spectrophotometer.[11] Average the duplicate values and subtract the absorbance of the reagent blank from the unknown and working standard.

NOTE: The order of performance of each of the steps in the procedure should remain constant. Up to 7 blood specimens may be measured in duplicate simultaneously. The procedure may be interrupted after diffusion by transferring the receiving rod to an empty diffusion bottle until it is convenient to complete the analysis. Sealed within the empty diffusion bottles the rods are safe from contamination. The ammonia concentration is not altered by such storage at room temperature for periods up to 6 hours. Longer periods of storage have not

[11] The Coleman Jr. spectrophotometer and the Coleman Electric Colorimeter, with a filter absorbing at 415 mμ, and 12 mm. cuvets, are satisfactory for this analysis. The Beckman DU spectrophotometer, with a 10 mm. absorption cell and a wavelength setting of 395 mμ, has also been used for this purpose (Checker G. D. D.).

been studied. To complete the analysis transfer the rods from the storage bottles into Nessler's solution.

As the color development follows Beer's law, the calculation of the ammonia concentration is based on the following ratios:

$$\frac{\text{Concentration of unknown sample in } \mu\text{g./ml.}}{\text{Concentration of standard } (3 \ \mu\text{g./ml.})} = \frac{\text{absorbance of unknown}}{\text{absorbance of standard}}$$

Calculate the ammonia concentration (corrected for the reagent blank) as follows:

$$\text{Unknown in } \mu\text{g. NH}_3\text{-N}/100 \text{ ml.} = \frac{\text{absorbance of unknown} \times 3}{\text{absorbance of standard}} \times 100$$

Blood Ammonia Levels

The mean normal fasting ammonia concentration of venous blood in 50 normal subjects was 102 ± 23 (S. D.) μg./100 ml., and ranged from 60 to 150. The fasting arterial ammonia level of 25 normal individuals averaged 106 ± 15 (S. D.) μg./100 ml. and ranged from 90 to 150.

Previous investigators have reported mean blood ammonia values in normal subjects which ranged from 11 to 204 μg./100 ml. (5, 6). This enormous range, which has directed valid criticism at all blood ammonia methods, has resulted from many modifications of existing methods including differences in the degree of alkalinization, the duration of diffusion, the interval between shedding of blood and its analysis, the anticoagulant employed, and many other factors.

In the resting state arterial ammonia levels are almost always slightly higher than venous levels (8, 17). After muscular exercise, however, the ammonia concentration of venous blood draining the active extremity surpasses arterial levels and results in a negative arteriovenous ammonia difference (13, 18, 19). Blood ammonia levels remain reasonably stable in normal individuals from day to day and are not greatly altered by diet. An increase in dietary protein in cirrhotic patients or patients with portacaval shunts, however, causes an increase in blood ammonia concentration roughly proportional to the degree of portal-systemic collateral circulation.

In cirrhotic patients with hepatic coma caused by gastrointestinal hemorrhage or the ingestion of ammonia-liberating nitrogenous substances, arterial ammonia may be increased to levels as high as 700 μg./100 ml. As arterial levels rise the arteriovenous ammonia difference tends to increase, resulting in relatively lower venous ammonia levels

(17, 20). Although the blood ammonia concentration tends to parallel the severity of impending hepatic coma, the correlation is far from perfect.

Discussion

The method described has been in continuous use in the Submitter's laboratory for 8 years and has proved extremely reliable and reproducible in the diagnosis, prognosis, and management of hepatic coma and gastrointestinal bleeding in patients with liver disease. In addition, it is the basis of the ammonia tolerance test which provides a reliable index of the presence and degree of portal-systemic shunting of blood in cirrhotic patients (21).

Standardization of the technique has shown that absorbance was proportional to ammonia concentration from 0 to 600 μg./100 ml.

The recovery of various concentrations of ammonium sulfate added to blood ranged from 84.2 to 104.8% and averaged 97.1% ± 5.3 (S. D.). Percentage recovery, which was based on the addition of 75 to 300 μg. of NH_3-N to 25 blood samples of varying ammonia concentration, was calculated using the amount of NH_3-N added plus the amount originally present as 100%.

The coefficient of variation, which was calculated for 5 to 8 simultaneous analyses of 15 different blood samples, ranged from 3.8 to 8.9% and averaged 6.1%. It was calculated according to the formula:

$$\text{Coefficient of variation} = \frac{\text{standard deviation}}{\text{mean}} \times 100$$

Although one would expect the Seligson–Hirahara ammonia method to give lower blood ammonia concentrations than the Conway method by virtue of less drastic alkalinization in the former method, normal blood ammonia levels average approximately 100 μg./100 ml. with the Seligson–Hirahara as compared to 45–50 μg./100 ml. with the Conway method (4). A careful comparison of the Seligson–Hirahara technique with the Conway method showed that although they are similar in adherence to Beer's law, in precision, and in recovery experiments, the former method is simpler technically and better suited to the performance of multiple determinations simultaneously. In addition, the colorimetric quantitation is faster and easier than the more tedious microtitration.

In 100 consecutive blood ammonia determinations performed simultaneously by both methods the values obtained with the Seligson–

Hirahara technique averaged 85.3 ± 23 (S. D.) $\mu g./100$ ml. higher than those with the Conway method. This difference remained relatively constant over a wide range of blood ammonia levels. Although there are many differences between the two methods which may account for the disparity in results, it appears to be caused by the relatively slower liberation of ammonia from blood than from the aqueous ammonium sulfate standard in the Conway method (22). The rates of diffusion of ammonia from both blood and aqueous solutions are similar with the Seligson–Hirahara technique (1, 22).

Whether ammonia actually exists in the free state in blood has long been a source of controversy. Conway found a sharp rise in ammonia concentration starting immediately after the withdrawal of blood, and concluded that there was no free ammonia in normal blood (4, 14). Some investigators have confirmed these observations (23, 24, 25), but others, employing the same or other methods (1, 12, 26, 27, 28) have been unable to do so. Conway also observed that the initial increase in blood ammonia concentration could be inhibited by maintenance of the carbon dioxide tension (4, 14). Although these observations have been confirmed by some investigators (29, 30, 31), they have not been substantiated by others (24, 32). Recent evidence has supported the concept that free ammonia does exist in circulating blood (33, 34, 35).

The *absolute concentration* of ammonia in blood, however, remains uncertain. Alkaline hydrolysis of proteins, and perhaps of other ammoniagenic substances, certainly contributes to the apparent ammonia concentration by both methods (36, 37). Nevertheless, there is good, although not perfect, correlation of the blood ammonia concentration with the degree of hepatic coma induced by nitrogenous substances (7, 38, 39). The correlation is improved by the use of arterial blood to avoid the artifactual effect of the arteriovenous ammonia difference (20, 39, 40). Furthermore, abnormalities in acid-base balance cause alterations in the pH gradient between the intra- and extracellular fluid compartments, which greatly influence the distribution and toxicity of ammonia (41, 42).

Despite its shortcomings, the blood ammonia determination is a valuable clinical and experimental tool in the management and study of liver disease. These determinations would be of greatest value if the results obtained in various laboratories were comparable. Numerous variations in technique including the type of anticoagulant used, the interval between the shedding of blood and its analysis, the duration of diffusion, the reagents employed in alkalinization, the composition of Nessler's reagent, and others, may influence the blood ammonia concentration. These many variables must be recognized and kept constant.

REFERENCES

1. Seligson, D., and Hirahara, K., The measurement of ammonia in whole blood, erythrocytes, and plasma. *J. Lab. Clin. Med.* **49**, 962–974 (1957).
2. Folin, O., and Denis, W., Protein metabolism from the standpoint of blood and tissue analysis. The origin and significance of the ammonia in the portal blood. *J. Biol. Chem.* **11**, 161–167 (1912).
3. Parnas, J. K., and Wagner, R., Über die Ausführung von Bestimmungen kleiner Stickstoffmengen nach Kjeldahl. *Biochem. Z.* **125**, 253–256 (1921).
4. Conway, E. J., Apparatus for the microdeterminations of certain volatile substances. IV. The blood ammonia with observations on normal human blood. *Biochem. J.* **29**, 27–55 (1935).
5. Singh, I. D., Barclay, J. A., and Cooke, W. T., Blood ammonia levels in relation to hepatic coma and administration of glutamic acid. *Lancet* **1**, 1004–1007 (1954).
6. Phillips, G. B., Schwartz, R., Gabuzda, G. J., and Davidson, C. S., The syndrome of impending hepatic coma in patients with cirrhosis of the liver given certain nitrogenous substances. *New Engl. J. Med.* **247**, 239–246 (1952).
7. Eiseman, B., Bakewell, W., and Clark, G., Studies in ammonia metabolism. I. Ammonia metabolism and glutamate therapy in hepatic coma. *Am. J. Med.* **20**, 890–895 (1956).
8. Tyor, M. P., and Wilson, W. P., Peripheral biochemical changes associated with the intravenous administration of ammonium salts in normal subjects. *J. Lab. Clin. Med.* **51**, 592–599 (1958).
9. Seligson, D., and Seligson, H., Microdiffusion method for the determination of nitrogen liberated as ammonia. *J. Lab. Clin. Med.* **38**, 324–330 (1951).
10. Conn, H. O., Effect of heparin on the blood ammonia determination. *New Engl. J. Med.* **262**, 1103–1107 (1960).
11. deGroote, J., and Vandenbroucke, J., Evaluation of blood ammonia after venipuncture. *Am. J. Digest. Diseases* **3**, 502–510 (1958).
12. Conn, H. O., Studies on the origin and significance of the blood ammonia. I. Effect of various anticoagulants on the blood ammonia determination. *Yale J. Biol. Med.* **35**, 171–184 (1962).
13. Allen, S. I., and Conn, H. O., Observations on the effect of exercise on blood ammonia concentration in man. *Yale J. Biol. Med.* **33**, 133–144 (1960).
14. Conway, E. J., and Cooke, R., Blood ammonia. *Biochem. J.* **33**, 457–478 (1939).
15. Rosenoer, V. M., The measurement of ammonia in whole blood. *J. Clin. Pathol.* **12**, 128–130 (1959).
16. Conn, H. O., and Kuljian, A. A., The preservation of blood by rapid freezing

for subsequent determination of blood ammonia. *J. Lab. Clin. Med.* **63,** 1033–1040 (1964).

17. Bessman, S. P., and Bradley, J. E., Uptake of ammonia by muscle; its implications in ammoniagenic coma. *New Engl. J. Med.* **253,** 1143–1147 (1955).

18. Parnas, J. K., Mozolowski, W., and Lewinski, W., Über den Ammoniakgehalt und die Ammoniakbildung im Blute. IX. Der Zusammenhang des Blutammoniaks mit der Muskelarbeit. *Biochem. Z.* **188,** 15–23 (1927).

19. Schwartz, A. E., Lawrence, W., Jr., and Roberts, K. E., Elevation of peripheral blood ammoina following muscular exercise. *Proc. Soc. Exptl. Biol. Med.* **98,** 548–550 (1958).

20. Bessman, S. P., and Bessman, A. N., The cerebral and peripheral uptake of ammonia in liver disease with an hypothesis for the mechanism of hepatic coma. *J. Clin. Invest.* **34,** 622–628 (1955).

21. Conn, H. O., Ammonia tolerance as an index of portal-systemic collateral circulation in cirrhosis. *Gastroenterology* **41,** 97–106 (1961).

22. Conn, H. O., A critical comparison of the Conway and Seligson-Hirahara blood ammonia methods. To be published.

23. Koprowski, H., and Uninski, H., Ammonia content of canine blood after oral administration of ammonium salts and ammonia. *Biochem. J.* **33,** 747–753 (1939).

24. White, L. P., Phear, E. A., Summerskill, W. H. J., and Sherlock, S., Ammonium tolerance in liver disease. Observations based on catheterization of the hepatic veins. *J. Clin. Invest.* **34,** 158–168 (1955).

25. Varay, A., Crosnier, J., and Masson, M., Étude critique de l'ammoniémie au cours des affections hépatiques. *Arch. Maladies App. Digest et Maladies Nutrition* **45,** 5–31 (1956).

26. Calkins, W. G., The blood ammonia in normal persons. *J. Lab. Clin. Med.* **47,** 343–348 (1956).

27. Hulme, W. A., and Cooper, A. C., Blood ammonium estimation as a routine laboratory procedure. *J. Med. Lab. Technol.* **13,** 543-547 (1956).

28. Merchant, A. C., Goldberger, R., and Barker, H. G., Preservation of blood for ammonia analysis. *J. Lab. Clin. Med.* **55,** 790–795 (1960).

29. Nathan, D. G., and Rodkey, F. L., A colorimetric procedure for the determination of blood ammonia. *J. Lab. Clin. Med.* **49,** 779–785 (1957).

30. Singh, I. D., Barclay, J. A., and Cooke, W. T., Hepatic coma. *Lancet* **2,** 335–336 (1954).

31. Strehler, E., Haas, J., and Rupp, F., Der Einfluss von CO_2 auf den Ammoniakgehalt des Blutes *in vitro. Biochem. Z.* **313,** 170–173 (1942).

32. Conn, H. O., Unpublished observations.

33. Jacquez, J. A., Poppell, J. W., and Jeltsch, R., Partial pressure of ammonia in alveolar air. *Science* **129,** 269–270 (1959).

34. Robin, E. D., Travis, D. H., Bromberg, P. A., Forkner, C. E., Jr., and Tyler, J. M., Ammonia excretion by mammalian lung. *Science* **129,** 270–271 (1959).

35. Bromberg, P. A., Robin, E. D., and Forkner, C. E., Jr., The existence of ammonia in blood *in vivo* with observations on the NH_4–NH_3 system. *J. Clin. Invest.* **39,** 332–341 (1960).

36. Reinhold, J. G., and Chung, C., Formation of artifactual ammonia in blood

by action of alkali: its significance for the measurement of blood ammonia. *Clin. Chem.* **7**, 54–69 (1961).

37. Reif, A. E., The ammonia content of blood and plasma. *Anal. Biochem.* **1**, 351–370 (1960).
38. McDermott, W. V., Jr., Metabolism and toxicity of ammonia. *New Engl. J. Med.* **257**, 1076–1081 (1957).
39. Stahl, J., Studies of the blood ammonia in liver disease: its diagnostic, prognostic and therapeutic significance. *Ann. Internal Med.* **58**, 1–24 (1963).
40. Fahey, J. L., Nathans, D., and Rairigh, D., Effect of L-arginine on elevated blood ammonia levels in man. *Am. J. Med.* **23**, 860–869 (1957).
41. Stabenau, J. R., Warren, K. S., and Rall, D. P., The role of pH gradient in the distribution of ammonia between blood and cerebrospinal fluid, brain and muscle. *J. Clin. Invest.* **38**, 373–383 (1959).
42. Warren, K. S., Iber, F. L., Dölle, W., and Sherlock, S., Effect of alterations in blood pH on distribution of ammonia from blood to cerebrospinal fluid in patients in hepatic coma. *J. Lab. Clin. Med.* **56**, 687–694 (1960).

BILIRUBIN (MODIFIED JENDRASSIK AND GROF)—Provisional

Submitted by: S. Raymond Gambino, Englewood Hospital, Englewood, New Jersey; and Columbia-Presbyterian Medical Center, New York, New York

Checked by: Walter R. C. Golden, Stamford Hospital, Stamford, Connecticut
Jesse F. Goodwin, Wayne State University, Children's Hospital of Michigan, Detroit, Michigan
A. William Smith, Clinical Laboratory of Las Vegas, Las Vegas, Nevada

Introduction

Ehrlich (1) described the coupling of bilirubin with diazotized sulfanilic acid to form a red pigment in neutral solutions and a blue pigment in strongly acid or alkaline solutions. Van den Bergh and Snapper (2) used this color reaction for quantitative measurements of serum bilirubin. Later, van den Bergh and Muller (3) discovered the "accelerator" effect of alcohol on the coupling reaction. Adler and Strauss (4) found acceleration with caffeine-sodium benzoate.

Jendrassik and Grof (5) combined caffeine-sodium benzoate with sodium acetate, diazotized with 0.5% (w/v) sulfanilic acid, and formed alkaline azobilirubin at pH 13.4. With (6), Fog (7), Nosslin (8), and Michaëlsson (9) have made detailed and favorable studies of this method. Schellong and Wende (10) and Schellong (11) reported favorably on a micro modification for pediatric use. The bilirubin method presented here is similar to Jendrassik and Grof's original method (5) except for proportional decreases in all volumes.

Principles

Serum or plasma is added to a solution of sodium acetate and caffeine-sodium benzoate. The sodium acetate buffers the pH of the diazo reaction, while the caffeine-sodium benzoate accelerates the coupling of bilirubin with diazotized sulfanilic acid. The azobilirubin color develops within 10 minutes.

55

A strong alkaline solution is then added to convert the pink azobilirubin to blue azobilirubin which is measured spectrophotometrically. Conversion to the blue pigment increases specificity by shifting the absorbancy maximum to 600 mμ. At 600 mμ, nonbilirubin yellow pigments and other red and brown pigments have negligible absorptivities. The final color appears green because blue alkaline azobilirubin is mixed with yellow pigment derived from a reaction between caffeine and sulfanilic acid. Jendrassik and Grof selected caffeine-sodium benzoate instead of alcohol to avoid protein precipitation. They chose alkaline instead of acid azobilirubin for the same reason.

Reagents

NOTE: All reagents should be of reagent grade quality unless otherwise specified.

1. Caffeine mixture. Add 50 g. of caffeine, purified alkaloid $C_{18}H_{10}N_4O_2$, 75 g. of sodium benzoate, C_6H_5COONa (USP), and 125 g. of sodium acetate, granular $CH_3COONa \cdot 3H_2O$, to distilled water at 50° to 60°C. and bring to 1 l. when cool. The reagent is stable for at least 6 months at room temperature when stored in glass or polyethylene bottles.

2. Diazo I. Add 5.0 g. of sulfanilic acid, $C_6H_4NH_2SO_3H \cdot H_2O$, and 15.0 ml. of concentrated hydrochloric acid, to distilled water and bring to 1 l. The reagent is stable for at least 6 months at room temperature when stored in glass or polyethylene bottles.

3. Diazo II. Add 500 mg. of sodium nitrite, $NaNO_2$, to distilled water and bring to 100 ml. This reagent is stable up to 2 weeks in a stoppered volumetric flask at 4° to 6°C.

NOTE: The Submitter uses preweighed samples of *dry* sodium nitrite. The dry sodium nitrite crystals have been stable at 20° to 30°C. for at least 2 years when kept in glass vials with cork stoppers.

4. Diazo reagent. Mix 10.0 ml. of Diazo I with 0.25 ml. of Diazo II. Use the reagent within 30 minutes of preparation.

5. Alkaline mixture. Add 100 g. of sodium hydroxide, NaOH, and 350 g. of potassium sodium tartrate, $KNaC_4H_4O_6 \cdot 4H_2O$, to distilled water and dilute to 1 l. This solution is stable for at least 6 months at room temperature when stored in glass or polyethylene bottles.

Apparatus

1. Spectrophotometer.[1]

2. Cuvets. 12 × 75 mm. round cuvets,[2] or 1.000 cm. precision rectangular silica cuvets.[3]

3. Pipets. Folin-Oswald pipets for 0.5 ml. and 1.0 ml. volumes. To Contain (T.C.) micropipets accurate to ± 0.5% (maximum error) for all volumes of 250 µl. or less.

Specimens

Serum or heparinized plasma can be used (9, 12, 13, 14).

NOTE: The Submitter has tested 10 serum versus plasma pairs and found no difference.

The specimens must be protected from daylight (15, 16, 17), and hemolysis should be avoided. However, the Jendrassik and Grof method is not significantly affected by hemoglobin levels up to 500 mg./100 ml. (see Discussion).

Procedure for Total Bilirubin

1. Add 1.00 ml. of caffeine mixture to each of two 12 × 75 mm. round cuvets (unknown and blank).

2. Add 50.0 µl. of serum or plasma to each cuvet followed by 200 µl. of distilled water. If the bilirubin is expected to be less than 5 mg./100 ml., more accurate results will be obtained by using 250 µl. of serum or plasma thereby eliminating the water dilution.

NOTE: The Jendrassik and Grof method is insensitive to a fiftyfold change in protein concentration (7). Therefore, a wide range of serum dilutions is avail-

---

[1] The Submitter and Checkers used the following instruments: Coleman Jr. model 6A, Beckman model B, Beckman model DU, and Bausch and Lomb Spectronic 20. The wavelength setting of the Beckman model B was calibrated with a mercury vapor lamp and the absorptivity response was checked with a Gilford model 202 optical density standard (Gilford Instrument Laboratories, Inc., Oberlin, Ohio). The wavelength setting of the Coleman model 6A was checked with a Coleman 6-400 didymium filter (Coleman Instruments, Inc., Maywood, Illinois).

[2] Coleman #6-308A

[3] Beckman #75170.

able. If special cuvets are used, e.g. 10 mm. I.D. perfect round cuvets[4] with spacers,[5] or Bessey-Lowry microcuvets,[6] then as little as 5.0 μl. of jaundiced serum can be used. All other volumes would be reduced in proportion.

3. Add 250 μl. of fresh diazo reagent to the unknown and mix.

4. Add 250 μl. of Diazo I to the blank and mix.

5. Wait 10 minutes.

NOTE: With (6) found that 90% of the color develops within 5 minutes. Maximum color is always reached within 10 minutes and is unchanged after 30 minutes.

6. Add 0.50 ml. of alkaline mixture to each cuvet and mix.

7. Within 30 minutes read the absorbance of the unknown at 600 mμ. with the blank at zero absorbance.

NOTE: Direct reacting bilirubin. This volume does not recommend a specific method for direct reacting bilirubin. However, for those laboratories requiring some estimate of bilirubin glucuronide, a tentative method used by the Submitter is given. This method is modified from Nosslin (8).
1. Add 1.00 ml. of 0.05 N hydrochloric acid to a 12 × 75 mm. round cuvet. A new blank is not required. Use the same blank used in the total bilirubin procedure.
2. Add 50.0 μl. of serum followed by 150 μl. of water and mix.
3. Add 250 μl. of fresh diazo reagent and mix.
4. Exactly 1 minute later add 50 μl. of 4% ascorbic acid (200 mg. in 5 ml. of distilled water) and immediately thereafter add 0.50 ml. of alkaline mixture and mix. Read in the same way as total bilirubin. The results obtained with this tentative method are similar to those obtained with the method of Ducci and Watson (18).

Standardization

This method should be standardized with a bilirubin-in-serum standard (see p. 76). Prepare a stock bilirubin-in-serum standard equal to 25.0 mg./100 ml. This stock may subsequently be diluted with water, saline, or serum diluent to obtain concentrations equal to 5.0, 10.0, 15.0, and 20.0 mg./100 ml. All dilutions must be made just prior to use. A pure aqueous standard cannot be used. Some protein must be present for maximum azobilirubin color, but the exact amount of protein is not critical (see Discussion).

The stock standard and its dilutions are used in 50.0 μl. amounts

[4] Scientific Products #68533A. Scientific Products, Evanston, Illinois (sole distributors).

[5] Scientific Products #68533B. Ibid.

[6] Pyrocell Manufacturing Co., Westwood, New Jersey.

in the standard total bilirubin procedure. The calibration curve is linear with the instruments used.[7]

NOTE: Several precautions must be observed when applying the committee recommendation on a uniform bilirubin standard. The standard should be made with human serum or plasma that is less than 4 hours old and that has been protected from light and heat. Many samples of older, heated, or irradiated sera caused slight to severe suppression of azobilirubin color when tested by the Submitter in four variations of the Malloy-Evelyn method (9, 18, 19, 20), the Powell-O'Hagan method (16), and the Jendrassik and Grof method. Michaëlsson (9) and Schellong (21) also recommended fresh serum. The standard must be protected from light. The pH of the standard should be brought to 7.3–7.5 (21, 22). When bilirubin crystals are dissolved in sodium carbonate there should not be any brown tint (23), and when the same crystals are dissolved in chloroform there should not be any brown tint. If 5 vol. of chloroform-bilirubin solution are washed with 1 vol. of 5% (w/v) sodium bicarbonate, the bicarbonate wash should be free of dark brown pigment (23).

NOTE: With (24) and Michaëlsson (9) have questioned the validity of measuring the molar absorptivity of bilirubin in chloroform for the control of purity. Both believe standards must also be defined in terms of azobilirubin molar absorptivity.

Properly prepared and controlled lyophilized reference samples will aid greatly in making the results of bilirubin analyses more uniform throughout the world (21).

Discussion

PRECISION

Michaëlsson (9) found a coefficient of variation (C.V.) of ± 2 to 3% for standards in fresh serum ranging from 5 to 25 mg./100 ml.

NOTE: The Submitter analyzed two control samples on 10 consecutive days. The C.V. of a 3 mg./100 ml. sample was ± 1.8%, and the C.V. of an 18 mg./100 ml. sample was ± 2.1%.

HEMOLYSIS

Hemolysis should be avoided, but moderate degrees of hemolysis do not present a problem. Michaelsson (9) found only 2 to 3% color loss with 280 mg./100 ml. of hemoglobin, and only 6 to 8% loss with 1400 mg./100 ml. of hemoglobin.

[7] Sample absorbancies with the Coleman Jr. model 6A spectrophotometer and 12 × 75 mm. round cuvets are: (5.0) 0.148, (10.0) 0.295, (15.0) 0.445, (20.0) 0.590, and (25.0) 0.740.

NOTE: The Submitter found a slight but variable effect when varying amounts of hemoglobin were added to 15 adult sera and 15 cord bloods. A maximum color loss of only 5% was found when hemoglobin was at 250 mg./100 ml., a maximum loss of 8.5% between 300 and 500 mg./100 ml., and a maximum loss of 15% between 500 and 1000 mg./100 ml. Sera with more than 250 mg./100 ml. of hemoglobin were cherry red. The greatest suppression was found with low bilirubin levels. Diluted serum samples with high bilirubin levels (greater than 5 mg./100 ml. before dilution) were less affected by hemoglobin interference. The required serum dilution decreased the final concentration of hemoglobin in the reaction mixture. In fact, diluted serum samples with as much as 2000 mg./100 ml. of hemoglobin before dilution showed only 3 to 6% suppression when read immediately.

NOTE: Checker (J. F. G.) tested the effect of hemoglobin on a bilirubin standard in albumin. He found 5% suppression at 200 mg./100 ml., 5% suppression at 400 mg./100 ml., and 11% suppression at 800 mg./100 ml. Checker (W. R. C. G.) found no suppression at 200 mg./100 ml., and 10% suppression at 400 mg./100 ml.

Michaëlsson (9) found that, if ascorbic acid was used in the total bilirubin procedure in the same concentration and volume as in the procedure for direct reacting bilirubin (see NOTE: Direct reacting bilirubin), the suppressive effect of as much as 1400 mg./100 ml. of hemoglobin was eliminated.

NOTE: The Submitter and Checker (J. F. G.) have confirmed the protection given by ascorbic acid. However, since hemoglobin has no significant effect below 250 mg./100 ml. and minimal effect up to 500 mg./100 ml., the Submitter does not recommend routine use of ascorbic acid in the total bilirubin procedure.

In addition to color suppression, hemoglobin causes fading of the azo color (9). Therefore, less or no suppression is seen when readings are taken immediately after adding the alkaline mixture rather than waiting up to 30 minutes.

NOTE: The Submitter has confirmed Michaëlsson's findings. For example, a sample containing 18 mg./100 ml. of bilirubin and 1000 mg./100 ml. of hemoglobin showed only 1% color loss compared to a nonhemoglobin-containing aliquot when read immediately. When read 30 minutes later, the hemoglobin-containing aliquot showed 12% total color loss while the nonhemoglobin-containing aliquot remained unchanged.

VOLUMES

Wide variation in proportional volumes of serum, caffeine mixture, sulfanilic acid and nitrite are without effect (7). However, the proportional volume of alkaline mixture cannot be increased without risking turbidity.

REACTION TEMPERATURE

The effect of variation in reaction temperature on the final azo color has not been reported.

NOTE: The Submitter has studied the rate of diazotization of sera at four temperatures: 0°, 25°, 37°, and 56°C. The alkaline mixture was added 1, 2, 4, 6, and 10 minutes after adding diazo reagent. The reaction was fastest at 25°C. At 56°C. the color decreased with increasing incubation time.

PROTEIN VERSUS AQUEOUS STANDARDS

The azobilirubin color is more intense when protein is present. Bilirubin standards in protein solutions give about 10% more azobilirubin color than standards in weak alkali (6, 8). However, a fifty-fold variation in protein concentration has no significant effect on azobilirubin color (7). Therefore, it is not necessary to keep the protein concentration constant.

NOTE: Michaëlsson (9) found no effect when fresh serum was diluted tenfold, and the Submitter checked a twelvefold dilution of twenty fresh serum specimens and found no effect. Checker (W. R. C. G.) uses 5% (w/v) albumin in place of water when diluting jaundiced specimens in order to keep the protein concentration constant. The Submitter, however, does not favor the routine use of albumin as it is not necessary and may suppress color formation. Some lots of commercial parenteral albumin tested by the Submitter suppressed azo color even though free of sodium azide. The Submitter did find reconstituted crystalized albumin to be a satisfactory diluent when stored at minus 20°C. after reconstitution with distilled water.

Jendrassik and Grof (5), With (6), Fog (7), Nosslin, (8), and Michaëlsson (9), found 99 to 101% recovery of bilirubin added to native serum or plasma. Michaëlsson (9) found a linear relationship between concentration and absorptivity when checked to a level of 25 mg./100 ml.

NOTE: The Submitter has confirmed the linear relationship to 30 mg./100 ml. Higher levels were not checked.

Standards in chloroform cannot be used in the Jendrassik and Grof procedure because no chloroform-miscible solvent is used.

AZOBILIRUBIN MOLAR ABSORPTIVITY

Michaëlsson (9) and With (24) recommend defining bilirubin standards in terms of azobilirubin molar absorptivity. Both authors question the validity of measuring the molar absorptivity of bilirubin in chloroform for the control of purity and standardization. The maximum molar absorptivity of alkaline azobilirubin was calculated by the Submitter to be 73,000.[8] Nosslin (8) found a molar absorptivity of 70,000.

NOTE: The maximum azobilirubin molar absorptivity of 73,000 was obtained by the Submitter with specially purified (23) bilirubin whose chloroform molar absorptivity was 63,500. When commercially available bilirubin with a chloroform molar absorptivity of 60,800 was used, the azobilirubin molar absorptivity was 70,000.

COMPARISON WITH OTHER METHODS

Watson (25) compared the Jendrassik and Grof method (5) with that of Lathe and Ruthven (20). He found 4 to 8% lower values with the Jendrassik and Grof method. On the other hand, Michaëlsson (9) compared the Jendrassik and Grof method with three variations of the Malloy-Evelyn method. The Jendrassik and Grof method gave slightly higher values than the classic Malloy-Evelyn procedure when bilirubin was in the 15 to 25 mg./100 ml. range. The Jendrassik and Grof method gave slightly lower values when compared with the Malloy-Evelyn procedure read at 60 minutes instead of the usual 30 minutes. Finally, excellent correlation was obtained when serum was diluted 1 plus 30 in the first step of the Malloy-Evelyn procedure.

[8] Recrystalized bilirubin (23) was dissolved in sodium carbonate in the dark and then added to fresh serum. The azobilirubin color was measured in 1.000 cm. cuvets with a Beckman model B, slit width 0.02 mm., and wavelength of 600 mμ. The original Jendrassik and Grof (5) reagent volumes were used so that the final volume was 5.1 ml. The molecular weight of bilirubin (584) and not azobilirubin was used in the calculation. The calculation was made as follows:

$$\text{MA} \frac{600 \text{ m}\mu}{1 \text{ cm.}} = \frac{\text{MW} \times (\text{A/cm.}) \times \text{final volume in ml.}}{\text{Total amount of bilirubin in mg.}}$$

where:

$$\text{MA} \frac{600 \text{ m}\mu}{1 \text{ cm.}} = \frac{\text{Molar absorptivity at 600 m}\mu \text{ for 1 cm.}}{\text{light path.}}$$

$$\text{MW} = \text{Molecular weight of bilirubin.}$$

$$\text{A/cm.} = \text{Absorptivity per cm. of light path.}$$

Then:

$$\text{MA} \frac{600 \text{ m}\mu}{1 \text{ cm.}} = \frac{584 \times 0.490 \times 5.1}{0.02}$$

$$\text{MA} \frac{600 \text{ m}\mu}{1 \text{ cm.}} = 73,000.$$

BILIRUBIN (MODIFIED JENDRASSIK AND GROF)—Provisional   63

NOTE: The Submitter has compared 34 patient samples (30 adults and 4 infants) using the Jendrassik and Grof method (5) and the Ducci-Watson method (18). Each method was calibrated with an appropriately diluted standard in fresh serum. No differences were found.

NOTE: Checker (J. F. G.) measured 12 specimens by the Jendrassik and Grof method (9), the Malloy-Evelyn method (19), and the Meites-Hogg spectrophotometric method (26). He found a high degree of correlation among the results of the three methods.

NORMAL VALUES

One hundred two blood donors ranged from 0.15 to 1.2 mg./100 ml. (8). (See p. 68 for a discussion of bilirubin levels in infants.)

REFERENCES

1. Ehrlich, P., Sulfodiazobenzol, ein Reagens auf Bilirubin. *Centr. Klin. Med.* **4**, 721–723 (1883).
2. van den Bergh, A. A. H., and Snapper, J., Die Farbstoffe des Blutserums. *Deut. Arch. Klin. Med.* **110**, 540–561 (1913).
3. van den Bergh, A. A. H., and Muller, P., Über eine direkte und eine indirekte Diazoreaktion auf Bilirubin. *Biochem. Z.* **77**, 90–103 (1916).
4. Adler, E., and Strauss, L., Beitrag zum Mechanismus der Bilirubinreaktion im Blut. *Klin. Wochschr.* **1**, 2285–2286 (1922).
5. Jendrassik, L., and Grof, P., Vereinfachte photometrische Methoden zur Bestimmung des Blutbilirubins. *Biochem. Z.* **297**, 81–89 (1938).
6. With, T. K., The diazo reaction of bilirubin. *Acta Physiol. Scand.* **10**, 181–192 (1945).
7. Fog, J., Determination of bilirubin in serum as alkaline "azobilirubin." *Scand. J. Clin. Lab. Invest.* **10**, 241–245 (1958).
8. Nosslin, B., The direct diazo reaction of bile pigments in serum. *Scand. J. Clin. Lab. Invest.* **12**, Suppl. 49, 1–176 (1960).
9. Michaëlsson, M., Bilirubin determination in serum and urine. *Scand. J. Clin. Lab. Invest.* **13**, Suppl. 56, 1–79 (1961).
10. Schellong, G., and Wende, U., Mikromethode zur Bestimmung des Serumbilirubins aus Kapillarblut bei Neugeborenen. *Arch. Kinderheilk.* **162**, 126–135 (1960).
11. Schellong, G., Mikromethode zur Bestimmung des Gesamtbilirubins bei Neugeborenen. *Aerztl. Lab.* **4**, 110–113 (1963).
12. Watson, D., Analytic methods for bilirubin in blood plasma. *Clin. Chem.* **7**, 603–625 (1961).
13. Jacobi, M., Finkelstein, R., and Kurlen, R., Serum and plasma bilirubin. A comparative study of one hundred cases. *Arch. Internal Med.* **47**, 759–763 (1931).
14. Bröchner-Mortensen, K., Über Bilirubinbelastung als Leberfunktionsprobe. *Acta Med. Scand.* **85**, 1–32 (1935).

15. Cremer, R. J., Perryman, P. W., Richards, D. H., and Holbrook, B., Photo-sensitivity of serum bilirubin. *Biochem. J.* **66**, 60P (1957).
16. O'Hagan, J. E., Hamilton, T., Le Breton, E. G., and Shaw, A. E., Human serum bilirubin. An immediate method of determination and its application to the establishment of normal values. *Clin. Chem.* **3**, 609–623 (1957).
17. Sims, F. H., and Horn, C., Some observations on Powell's method for determination of serum bilirubin. *Am. J. Clin. Pathol.* **29**, 412–417 (1958).
18. Ducci, H., and Watson, C. J., The quantitative determination of the serum bilirubin with special reference to the prompt reacting and the chloroform soluble types. *J. Lab. Clin. Med.*, **30**, 293–300 (1945).
19. Malloy, H. T., and Evelyn, K. A., The determination of bilirubin with the photoelectric colorimeter. *J. Biol. Chem.* **119**, 481–490 (1937).
20. Lathe, G. H., and Ruthven, C. R. J., Factors affecting the rate of coupling of bilirubin and conjugated bilirubin in the van den Bergh reaction. *J. Clin. Pathol.* **11**, 155–161 (1958).
21. Schellong, G., Methods of serum bilirubin estimation. Problems associated with standardization and the introduction of an international standard bilirubin solution. *German Med. Monthly* **3**, 274–280 (1963).
22. Trainer, T. D., Preparation of a stable bilirubin standard. Summary Reports, Issue #9, p. 1, April, 1963. Published by Commission on Continuing Education of ASCP, Chicago, Illinois.
23. Henry, R. J., Jacobs, S. L., and Chiamori, N., Studies on the determination of bile pigments. I. Standard of purity for bilirubin. *Clin. Chem.* **6**, 529–536 (1960).
24. With, T. K., Neonatal jaundice. *Lancet* **2**, 618 (1962).
25. Watson, D., and Rogers, J. A., A study of six representative methods of plasma bilirubin analysis. *J. Clin. Pathol.* **14**, 271–278 (1961).
26. Meites, S., and Hogg, C. K., Direct spectrophotometry of total serum bilirubin in the newborn. *Clin. Chem.* **6**, 421–428 (1960).

BILIRUBIN (MODIFIED MALLOY AND EVELYN)—
Provisional

Submitted by: RODERICK P. MACDONALD, Harper Hospital, Detroit, Michigan
Checked by: RICHARD J. HENRY and S. L. JACOBS, Bio-Science Laboratories, Los Angeles, California
FRANK A. IBBOTT, University of Colorado Medical Center, Denver, Colorado
ALAN MATHER, The Memorial Hospital, Wilmington, Delaware

Introduction

In 1913, van den Bergh first applied Ehrlich's diazo reaction to the determination of serum bilirubin (1). He thus began a sequence of events resulting in more disappointment, dissention, and disenchantment than has occurred in any other area of clinical biochemistry. One salient fact persisted throughout all these years: bilirubin is an important blood constituent in clinical medicine. While research biochemists engaged in a dispute over the nature of *direct* and *indirect* reacting bilirubin, clinical laboratories calmly performed these analyses, providing information useful in the diagnosis and treatment of disease.

The current area of conflict lies in methods for analyzing serum bilirubin. Criteria for acceptable bilirubin standards have been provided (see p. 75). The method described here, though not overcoming the objections in bilirubin methodology, does provide a relatively simple and reliable procedure for measuring bilirubin in very small samples obtained from newborn babies and very young children.

A bilirubin method previously appeared in Volume 1 of "Standard Methods of Clinical Chemistry" (2). The present method uses less sample (vital in the case of newborn babies) and gives higher sensitivity through the use of modified diazo reagents. It also requires a shorter reaction time.

Principle

The bilirubin in serum is diazotized with Ehrlich's reagent to form azobilirubin (Scheme 1), following the principle of van den Bergh.

65

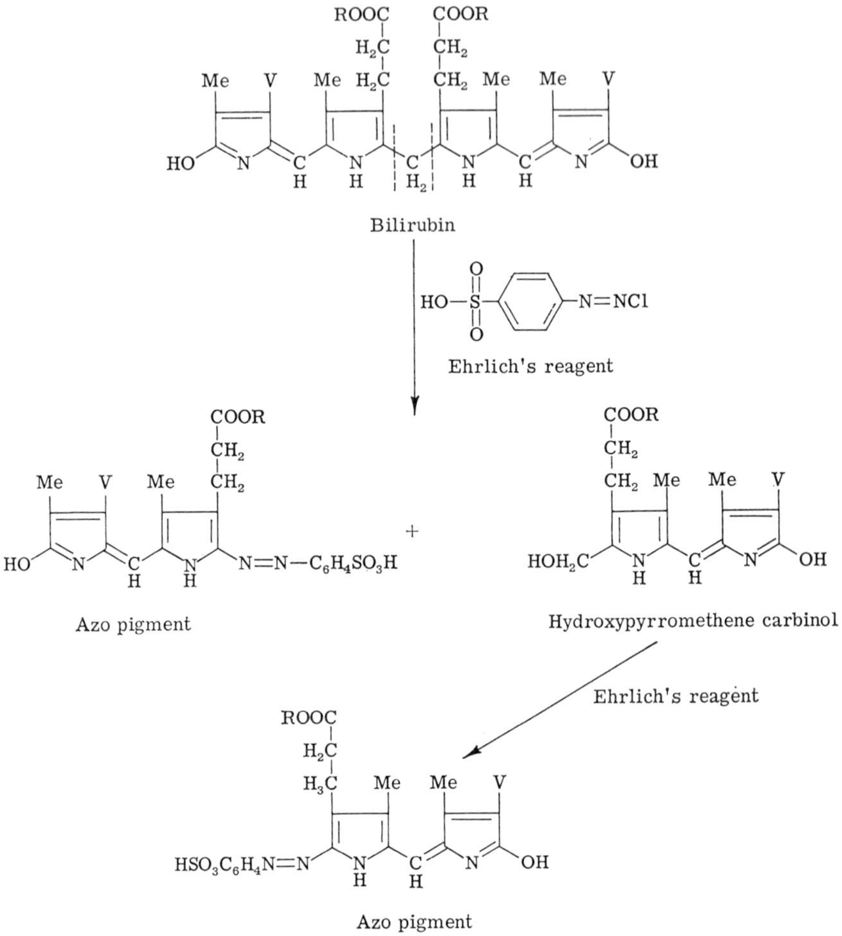

Bilirubin

Ehrlich's reagent

Azo pigment

Hydroxypyrromethene carbinol

Ehrlich's reagent

Azo pigment

V = Vinyl; Me = Methyl.
Conjugated bilirubin, R = H; Bilirubin glucuronide, R = Glucuronyl.

SCHEME 1

Malloy and Evelyn introduced the use of 50% methanol in the procedure, and eliminated the necessity of precipitating the proteins (3). This ultramicro modification (4) uses the reagent concentrations suggested by Meites (5, 6). Increased concentrations of hydrochloric acid, sulfanilic acid, and sodium nitrite increase the absorbance of azobilirubin and accelerate the reaction time to less than 10 minutes. The color developed is stable at least 2 hours at bilirubin concentrations of less than 15 mg./100 ml.

Reagents

1. *Diazo A reagent.* Dissolve 5 g. of sulfanilic acid in 60 ml. concentrated hydrochloric acid, mix and dilute to 1 l. with water. This solution is stable indefinitely.

2. *Sodium nitrite, 20% w/v.* Dissolve 20 g. of reagent sodium nitrite, $NaNO_2$, in water and dilute to 100 ml. Store under refrigeration in a brown glass-stoppered bottle. This solution should keep indefinitely, but discard if it becomes tinged with yellow.

3. *Diazo B reagent.* Dilute the 20% sodium nitrite solution 1:10 (e.g., 0.5 ml. + 4.5 ml. water). Prepare the reagent daily.

4. *Diazo blank reagent.* Dilute 60 ml. of concentrated hydrochloric acid to 1 l. with water. This solution is stable indefinitely.

5. *Diazo reagent.* Mix 0.15 ml. of reagent B and 5 ml. of reagent A. This reagent is stable about 30 minutes.

6. *Methyl alcohol, absolute, A.C.S. grade.*

Procedure

1. Pipet 250 μl. of water into the bottom of a 7 × 70 mm. test tube. Add 25 μl. of serum and mix thoroughly by gently tapping. This is a serum dilution of 1 + 10.

Note: Rinse the pipet with water diluent if it is calibrated "to contain."

2. Pipet 100 μl. of methanol into each of two 7 × 70 mm. test tubes labeled Sample and Blank.

Note: Test tubes and cuvets should be covered to prevent evaporation. Parafilm is convenient for this purpose.

3. Pipet 20 μl. of *fresh* diazo reagent into the sample tube, and 20 μl of diazo blank into the blank tube.

4. Add 100 μl. of diluted serum to each. Mix and allow to stand at room temperature for at least 10 minutes.

5. Measure the absorbance of the sample and blank at 560 mμ. Set the instrument at 0 absorbance with distilled water.

Note: Spectral absorbance curves of azobilirubin in serum were run by Checker (F. A. I.) and the absorbance maximum at 560 mμ was confirmed.

Note: The Submitter uses 25 × 2.5 × 10 mm. Bessey-Lowry cuvets which require a volume of 0.075 ml.[1]

[1] These cuvets and diaphragm are available from the Pyrocell Mfg. Co., 91 Carver Ave., Westwood, New Jersey. Such attachments are also available for other instruments.

NOTE: To determine "direct" reacting bilirubin (bilirubin glucuronide), follow the same procedure except substitute water for methanol, and read the colored solution in exactly 5 minutes. Some workers prefer to read the colored solution after 1 minute. Hogg and Meites have shown that the increased absorbance obtained after the longer period is due primarily to more direct reacting bilirubin being coupled (6). Definitive studies of the optimum time will require a purer form of conjugated bilirubin than is presently available.

NOTE: Checker (A. M.) varied this technique by scaling it up to micro use (0.1 ml. serum). He used the 1 + 10 serum, dilution and applied the 5:1:5 (v/v/v) ratio of methanol, reagent, and diluted serum. Linearity was quite good over the range examined (up to 25 mg./100 ml.). The slope varied by 1.5% between 5 and 25 mg./100 ml. The effect of inverting the order of serum dilution:reagent:methanol showed only slight lowering of the absorbance, and the slope varied 2% over this range. This Checker emphasized the importance of the minimum 10 mm. light path, a point which has also been stressed by the Submitter (7). Checkers (R. J. H. and S. L. J.) recommended a twofold increase of all volumes to permit use of a slightly larger capacity cuvet.

Calibration

Prepare bilirubin standards of 2.5, 5.0, 7.5, 10.0, 15.0, 20.0, and 25.0 mg./100 ml. following the directions given on page 76. Prepare a calibration curve to determine the extent of linearity for the instrument being used. Daily analyses should be checked with a control serum which meets the specifications on page 76.[2]

Normal Values

INFANTS

Approximate maximum levels of total bilirubin for newborns and infants are given in the tabulation (8).

Age	Premature (mg./100 ml.)	Full term (mg./100 ml.)
Up to 24 hours	Up to 8	5
Up to 48 hours	Up to 12	9
Day 3–5	Up to 24	12
After 1 month	0–0.3 (conjugated)	0.1–0.7 (unconjugated)

Behrendt (9) stresses that there is no critical bilirubin level for development of kernicterus. It may be seen in infants whose bilirubin

[2] Versatol A, A-Alternate, and Pediatric, prepared by Warner-Chilcott, Morris Plains, New Jersey, satisfies these specifications.

level never exceeds the range of 6–15 mg./100 ml. However, kernicterus is unlikely to occur when bilirubin concentration is below
20 mg./100 ml., and is very likely with levels of over 30 mg./100 ml.
Harris (10) studied 60 nonerythroblastotic premature infants and
found 73% had peak bilirubin levels below 15 mg./100 ml., and that
the largest number reached this peak on the fourth day of life. Ten
per cent had peak levels over 20 mg./100 ml., and these peaks
occurred at a later time. The highest bilirubin level of the group was
25.1 mg./100 ml. Serum bilirubin concentration was found to be related to birth weight; the higher the weight, the lower the peak bilirubin level. White infants weighing 1500 g. at birth had higher levels
than were obtained from Negro babies of the same weight. Female
babies had slightly lower bilirubin levels than male babies. Other
workers have obtained similar results.

Discussion

Numerous disadvantages of the Malloy-Evelyn diazo technique
have been reported (11, 12, 13). These include:

1. Sulfanilic acid must be diazotized immediately prior to use.

2. The long waiting period required for development of the color.

3. The van den Bergh reaction is not specific for bilirubin. Other
heme-derived pigments and unknown serum constituents influence
the reaction.

4. Serum protein turbidity may occur under certain conditions.

5. Although methanol is the best protein-releasing agent, it is more
likely to permit turbidity than other agents.

6. Large multiplier factors are required due to the relatively low
sensitivity of the reaction.

7. Diazo reactions may not produce consistent absorbances from
day to day.

8. Azobilirubin is a double-range pH indicator (14). Therefore,
control of the pH during the test procedure is essential.

9. Serum hemolysis decreases the amount of azobilirubin formed
in proportion to the amount of hemoglobin present.

These objections are not overwhelming as attested to by the usefulness of the Malloy-Evelyn procedure in thousands of laboratories for
many years. Therefore, the objective here is to provide a good micro
modification of this familiar procedure. No technique will give high
sensitivity at low concentrations of bilirubin and also be linear through

the range of pediatric interest without repetition of the test using a greater dilution of serum. Lack of good sensitivity for bilirubin concentrations under 1.0 mg./100 ml. may offend some chemists, but not those who are fully aware of the clinical significance of their test results. This procedure is recommended particularly for those who are performing their analyses for pediatric patients with either physiological or hemolytic jaundice.

This procedure has the following advantages:

1. It shows good linearity through the concentration range of pediatric interest (up to 25 mg. bilirubin/100 ml.).[3] The method will generally cover most bilirubin analyses without further dilution.

2. The required reaction time is only 10 minutes.

3. The azobilirubin color is stable for at least 2 hours for bilirubin levels of less than 15 mg./100 ml.

4. Turbidity has failed to be a serious problem after several years of using this method in the Submitter's laboratory. Further dilution of serum beyond that given in the procedure (e.g., a preliminary 1 + 1 dilution of the serum sample, then using this for the 1 + 10 dilution) may lead to erratic results when the bilirubin level is low (A. M.).

NOTE: This may be partially explained by the fact that the absorbance of the serum blanks did not dilute out proportionally. The 1 + 1 diluted serum substituted into the procedure as a preliminary step showed blanks of one-third to one-fourth of those for undiluted serum. This effect was observed only when the method was scaled upwards and a spectrophotometer used. It did not occur with a filter instrument. Checker (A. M.) believes this to be a turbidity effect, since the diazo reagent has some protective influence on serum turbidity. Turbidity may account for variance in the low and normal ranges of methanolic procedures; this is exacerbated with higher concentrations of sulfanilic acid in the reagent, and it is impossible to check whether the turbidities in the serum blank are equal to those in the diazo reaction. The Checker looked carefully for turbidity, and found visually preceptible quantities only occasionally. These correlated generally with higher blank values. The Submitter has discussed the problem of turbidity in ultramicro samples and warned of giving this proper attention when scaling down macro procedures (7).

[3] Sample absorbances are: (2.5) 0.110, (5.0) 0.220, (10) 0.455, and (20.0) 0.915 (Figures in parentheses are mg. bilirubin/100 ml. serum. The bilirubin used for these standards meets the requirements on p. 75.) These data were obtained on a Beckman Model DU spectrophotometer using Bessey-Lowry cuvets having a light path of 10 mm. In the experience of the Submitter and two Editors (S. M. and S. R. G.) calibration absorbances are very constant over long periods of time. However, a regular control program is still essential.

5. The Submitter has found this method to be relatively trouble-free and to give reliable results.

Incomplete coupling of the bilirubin may occur if the room temperature varies above or below 22–27°C. (5). Analysis should be performed immediately after drawing the specimen. O'Hagan studied the effect of bright sunlight for 1 hour on a specimen which initially contained 9.0 mg./100 ml. (15). After light exposure the value dropped to 4.8 mg./100 ml.

Hemolysis of the specimen should be avoided because of the interference of hemoglobin in the azo-coupling reaction (5). It is also important to avoid contaminating the reagent blank with the diazo reagent. If the same cuvet is used for both samples, the reagent blank should be read first.

REFERENCES

1. van den Bergh, A. A. H., and Snapper, J., Die Farbstoffe des Blutserums. *Deut. Arch. Klin. Med.* **110**, 540–561 (1913).
2. Kingsley, G. R., Getchell, G., and Schaffert, R. R., Bilirubin. *In* "Standard Methods of Clinical Chemistry" (M. Reiner, ed.), Vol. 1, pp. 11–15. Academic Press, New York, 1953.
3. Malloy, H. T., and Evelyn, K. A., The determination of bilirubin with the photoelectric colorimeter. *J. Biol. Chem.* **119**, 481–490 (1937).
4. Knights, E. M., Jr., MacDonald, R. P., and Ploompuu, J., "Ultramicro Methods for Clinical Laboratories," 2nd ed., pp. 60–65. Grune & Stratton, New York, 1962.
5. Meites, S., and Hogg, C. K., Studies on the use of the van den Bergh reagent for the determination of serum bilirubin. *Clin. Chem.* **5**, 470–478 (1959).
6. Hogg, C. K., and Meites, S., A modification of the Malloy and Evelyn procedure for the micro-determination of total serum bilirubin. *Am. J. Med. Technol.* **25**, 281–286 (1959).
7. MacDonald, R. P., Some theoretical considerations in adapting macro chemical procedures to an ultramicro scale. *Clin. Chem.* **8**, 450 (1962).
8. O'Brien, D., and Ibbott, F. A., "Laboratory Manual of Pediatric Micro- and Ultramicro-Biochemical Techniques," 3rd ed. Harper & Row, New York, 1962.
9. Behrendt, H., "Diagnostic Tests in Infants and Children," 2nd ed. Lea & Febiger, Philadelphia, Pennsylvania, 1962.
10. Harris, R. C., Peak levels of serum bilirubin in normal premature infants. *In* "Kernicterus" (A. Sass-Kortsäk, ed.), pp. 10–12. Univ. of Toronto Press, Toronto, Ontario, 1961.
11. Rand, R. N., and di Pasqua, A., A new diazo method for the determination of bilirubin. *Clin. Chem.* **8**, 570–578 (1962).

72 RODERICK P. MACDONALD

12. Mathew, A., Reliability of bilirubin determinations in icterus of the newborn infant. *Pediatrics* **26**, 350–354 (1960).
13. Michaëlsson, M., Bilirubin determination in serum and urine. *Scand. J. Clin. Lab. Invest.* **13**, Suppl. 56, 1–79 (1961).
14. Jirska, N., and Jirsova, V., Spectrophotometric behaviour of azobilirubin and azotaurobilirubin. *Clin. Chem.* **5**, 532–541 (1959).
15. O'Hagan, J. E., Hamilton, T., Le Breton, E. G., and Shaw, A. E., Human serum bilirubin. An immediate method of determination and its application to the establishment of normal values. *Clin. Chem.* **3**, 609–623 (1957).

Other Books, Articles, and Reviews[4]

General

16. Billing, B. H., Bile pigments in jaundice. *Advan. Clin. Chem.* **2**, 268–299 (1959).
17. Arias, I. M., The chemical basis of kernicterus. *Advan. Clin. Chem.* **3**, 35–82 (1960).
18. Sass-Kortsäk A., ed., "Kernicterus." Univ. of Toronto Press, Toronto, Ontario. 1961.

On the Nature of Bilirubin

19. Najjar, V. A., and Childs, B., The crystallization and properties of serum bilirubin. *J. Biol. Chem.* **204**, 359–366 (1953).
20. Cole, P. G., Lathe, G. H., and Billing, B. H., Separation of the bile pigments of serum, bile and urine. *Biochem. J.* **57**, 514–518 (1954).
21. Schmid, R., Direct reacting bilirubin, bilirubin glucuronide in serum, bile and urine. *Science* **124**, 76–77 (1956).
22. Schachter, D., Nature of the glucuronide in direct reading bilirubin. *Science* **126**, 507–508 (1957).
23. Billing, B. H., Cole, P. G., and Lathe, G. H., The excretion of bilirubin as a diglucuronide giving the direct van den Bergh reaction. *Biochem. J.* **65**, 774–784 (1957).
24. Billing, B. H., and Lathe, G. H., Bilirubin metabolism in jaundice. *Am. J. Med.* **24**, 111–121 (1958).
25. Schmid, R., Jeliu, G., and Gellis, S. S., Glucuronic acid and hyperbilirubinemia. *Science* **127**, 1512 (1958).
26. Schachter, D., Estimation of bilirubin mono- and diglucuronide in the plasma and urine of patients with nonhemolytic jaundice. *J. Lab. Clin. Med.* **54**, 557–562 (1959).
27. Schoenfield, L. J., Grindlay, J. H., Foulk, W. T., and Bollman, J. L. Identification of extrahepatic bilirubin monoglucuronide and its conversion to pigment 2 by isolated liver. *Proc. Soc. Expel. Biol Med.* **106**, 438–441 (1961).
28. Gregory, C. H., and Watson, C. J., Studies of conjugated bilirubin. I. Comparison of conventional fractional determination with chromatographic and

[4] Additional pertinent references may be found on pp. 63–64.

solvent partition methods for free and conjugated bilirubin. II. Problem of sulfates of bilirubin *in vivo* and *in vitro*. *J. Lab. Clin. Med.* **60,** 1–6, 17–30 (1962).

29. Broderson, R., Contribution to the identification of bile pigments in normal human serum. *Scand. J. Clin. Lab. Invest.* **14,** 517–527 (1962).

Kernicterus

30. Rapmund, G., Bowman, J. M., and Harris, R. C., Bilirubinemia in non-erythroblastotic premature infants. *Am. J. Diseases Children* **99,** 604–616 (1960).
31. Hugh-Jones, K., Slack, J., Simpson, K., Grossman, A., and Hsia, D. Y. Y., Clinical course of hyperbilirubinemia in premature infants. *New Engl. J. Med.* **263,** 1223–1229 (1960).
32. Waters, W. J., and Porter, E. G., Dye-binding capacity of serum albumin in hemolytic diseases of the newborn. *Am. J. Diseases Children* **102,** 807–814 (1961).
33. Watson, D., The amniotic fluid bile pigments in relation to haemolytic disease of the newborn. *Proc. Assoc. Clin. Biochemists* **2,** 81–83 (1962).
34. Reiner, M., and Thomas, J. L., Bilirubin studies on the serum of newborn infants, *Clin. Chem.* **8,** 278–283 (1962).

Eberleins's Solvent Partition Method

35. Eberlein, W. R., A simple solvent-partition method for measurement of free and conjugated bilirubin in serum. *Pediatrics,* **25,** 878–885 (1960).
36. Tisdale, W. A., and Welch, J., Study of serum bile pigments in jaundiced patients using the Eberlein solvent-partition method. *J. Lab. Clin. Med.* **59,** 956–962 (1962).

Direct Spectrophotometric Methods

37. Abelson, N. M., and Boggs, T. R., Jr., Plasma pigments in erythroblastosis. I. Spectrophotometric absorption patterns. *Pediatrics* **17,** 452–460 (1956).
38. White, D., Abu Haidar, G. A., and Reinhold, J. G., Spectrophotometric measurement of bilirubin concentrations in the serum of the newborn by the use of a microcapillary method. *Clin. Chem* **4,** 211–222 (1958).
39. Chiamori, N., Henry, R. J., and Golub, O. J., Studies on the determination of bilirubin and hemoglobin in serum. *Clin. Chim. Acta* **6,** 1–6 (1961).

Other Interesting Papers

40. Watson, D., and Rogers, J. A., A study of six representative methods for plasma bilirubin analysis. *J. Clin. Pathol.* **14,** 271–278 (1961).
41. Naumann, H. N., and Young, J. M., Comparative bilirubin levels in vitreous body, synovial fluid, cerebrospinal fluids, and serum after death. *Proc. Soc. Exptl. Biol. Med.* **105,** 70–72 (1960).
42. Berman, L. B., Lapham, L. W., and Pastore, E., Jaundice and xanthochromia of the spinal fluid. *J. Lab. Clin. Med.* **44,** 273–279 (1954).

74   RODERICK P. MACDONALD

43. Philpot, G. R., Bilirubin determination: interference by copper. *Proc. Assoc. Clin. Biochemists* **2**, 102–103 (1963).
44. Blondheim, S. H., Lathrop, D., and Zabriskie, J., Effect of light on the absorption spectrum of jaundiced serum. *J. Lab. Clin. Med.* **60**, 31–39 (1962).
45. McGann, C. J., and Carter, R. E., The effect of hemolysis on the van den Bergh reaction for serum bilirubin. *J. Pediat.* **57**, 199–203 (1960).
46. Bolt, R. J., Dillon, R. S., and Pollard, H. M., Interference with bilirubin excretion by a gall-bladder dye (bunamiodyl). *New Engl. J. Med.* **265**, 1043–1045 (1961).

RECOMMENDATION ON A UNIFORM
BILIRUBIN STANDARD*[1]

The management of hyperbilirubinemia in the newborn infant by exchange transfusion has been hampered by failure to agree consistently upon analytic data or bilirubin concentrations [Mather, A. *Pediatrics* **26**:350 (1960]. In order to eliminate one of the variables, a committee composed of representatives from the American Academy of Pediatrics, the College of American Pathologists, the American Association of Clinical Chemists, and the National Institutes of Health, recommends the following procedures for the establishment of a uniform bilirubin standard.

Acceptable Bilirubin

Since there are at present no generally accepted criteria of purity of bilirubin preparations and thus no basis for the definition of a primary bilirubin standard at this time, the Committee has had tentatively to accept the principle that increasing molar absorptivity at bilirubin's maximum in the visible region is an index of purity in the sample. On this basis the Committee has examined six highly purified commercial and two privately purified crystalline preparations selected as giving the highest and most consistent molar absorptivity in chloroform at 453 mμ. The results of this examination in triplicate by three separate laboratories give a 1-cm. molar absorptivity of 60,700 ± 800 mean ± standard deviation) at 453 mμ in chloroform at 25°C. The Committee recommends as acceptable a bilirubin giving an absorptivity between 59,100 and 62,300.

* Developed jointly by representatives of the American Academy of Pediatrics, the College of American Pathologists, the American Association of Clinical Chemists, and the National Institutes of Health. The American Association of Clinical Chemists was represented by a Subcommittee on Bilirubin of the Standards Committee. The committee consists of Frank Ibbott, Alan Mather, Willard Faulkner, Samuel Meites, and Richard Henry, Chairman. The College of American Pathologists (Prudential Plaza, Chicago 1, Ill.), is willing to certify bilirubin preparations as "acceptable" as defined in this recommendation.
[1] Reproduced from *Clin. Chem.* **8**, 405–407 (1962), with the permission of Hoeber Medical Division of Harper & Row, New York.

Standard Solutions for the Assay of Serum Bilirubin

In view of widely reported observations that serum proteins may affect the absorptivities of bilirubin and azobilirubin, and of the desirability of calibration of assay procedures under conditions as much as possible like those with which the analyzed specimen is to be handled, the Committee recommends that standard solutions for the assay of bilirubin in serum be made up in an aqueous protein medium. The choice obviously is between the use of artifically prepared buffered serum protein solutions and the use of pooled serum as the medium for the preparation and dilution of standard solutions. Although both have disadvantages, the latter was selected on the basis of its ready availability and probable superiority.

An "acceptable standard bilirubin solution" is prepared as follows: An accurately weighed quantity of an "acceptable bilirubin" is dissolved completely and as quickly as possible (less than 5 min.) in subdued light at room temperature in $M/10$ sodium carbonate solution, the quantity of the latter being selected to constitute 2 per cent of the final volume of the prepared standard.[2] The clear red solution is immediately diluted with an "acceptable serum diluent" to a final volume selected to give a desired final concentration of not less than 5 mg./100 ml. An "acceptable serum diluent," to be used both in preparation of this standard and in any subsequent dilutions for calibration purposes is defined tentatively as pooled serum having an absorbance of less than 0.100 at 414 mμ and 0.040 at 460 mμ at a dilution of 1:25 in 0.85% NaCl.

For calibration purposes, a concentrated standard may be diluted with the same "serum diluent" to appropriate concentrations covering adequately the range for which the assay procedure is designed. Standard and subdilutions as well as the "serum diluent" are to be treated in an analytical procedure in a manner identical with that for unknown serum specimens, and corrections for the color contributed by the serum diluent must be made. The absorbance (A) of a standard (read against its own undiazotized blank in any diazo procedure) is A_s. To correct for that part of A_s contributed by the serum diluent itself, the absorbance of serum diluent alone (read against its own

---

[2] Occasionally, bilirubin preparations are encountered which are difficult to dissolve in the carbonate. When this occurs, 0.05 N sodium hydroxide may replace the 0.10 M sodium carbonate, provided the resultant solution is used within 5 minutes (Richard J. Henry).

blank in diazo procedures) constitutes A_b. The "corrected A_s," -i.e., the absorbance equivalent to the weighed-in standard concentration, is $A_s - A_b$. If handled in the above manner, the nominal concentration of the standard may be taken as that of the weighed-in bilirubin.

Packaging and Preservation

Standards prepared in the above manner are not stable preparations and may be preserved for about a week at $-20°C$. For any commercial packaging of standards so prepared, suitable aliquots are to be lyophilized and sealed in ampoules or bottles. Shelf life of such preparations must be established and, if found to be stable for only a limited time, the expiration date must be indicated on the container label (current indications are that stability of such lyophilized material cannot be assumed). It is recommended that the lyophilized preparations be kept under refrigeration and that a reconstituted standard be used within 2 hours. The stated concentrations of such reconstituted standards must be based upon (1) the sum of the added bilirubin and that calculated to be contributed by the serum diluent, and (2) a carefully determined correction derived from the ratio of the prelyophilized and reconstituted volumes.

Calibration Using Preserved Standards

Because of the large variations obtained among methods, any assay procedure should be calibrated over the concentration range for which it yields linear results. With methods employing aqueous dilution of highly jaundiced sera, a suitable standard must first be prepared and subdiluted in serum to appropriate concentrations, and then diluted with water in a manner identical with that used for the specimens. For procedures which bring the concentration into the linear range by the use of micro aliquots, the various subdilutions in serum are sampled directly.

For preserved standards which are to be diluted in a "serum diluent" which may vary significantly from the original diluent, or for subdilutions from reconstituted lyophilized standards, the nominal (weighed-in) equivalent concentration cannot be used, and the value for total concentration—i.e., the sum of weighed and diluent bilirubin—must be used as the basis for calculation. Under these conditions, each dilution of standard in serum must be corrected for the corresponding dilution of the serum diluent; a 1:4 dilution of standard in serum must be corrected by ¾ of the determined serum bilirubin absorbance, a 1:2 dilution by ½ the serum absorbance, and so on.

With regard to the problem of standardization of measurements of conjugated bilirubins in serum, the Committee cannot provide any recommendation at the present time.

References

1. Recommendation on a uniform bilirubin standard. *Clin. Chem.* **8**, 405–407 (1962).
2. A uniform bilirubin standard. Recommendations of the College of American Pathologists Standards Committee. *Am. J. Clin. Pathol.* **39**, 90–91 (1963).
3. Recommendations on a uniform bilirubin standard. *Pediatrics* **31**, 878–880 (1963).

TOTAL AND FREE CHOLESTEROL*

Submitted by: BENNIE ZAK, Wayne State University College of Medicine, Department of Pathology, Detroit, Michigan

Checked by: MARTA CANCIO, VINCENTE BESARES-CASTRO, BENJAMIN CALDERÓN-TOMEI, and JOSÉ M. LEÓN, Medical Research Laboratory, Veterans Administration Hospital, San Juan, Puerto Rico

CLEAMOND D. ESKELSON, ROBERT FERRIS, and MARY CLARE HAVEN, Veterans Administration Hospital, Omaha, Nebraska

WILLIAM E. GLENNON, Heart Disease Epidemiology Study, Department of Health, Education and Welfare, Framingham, Massachusetts

ROBERT F. WITTER and VALETA K. MCDANIEL, Communicable Disease Center, Department of Health, Education and Welfare, Atlanta, Georgia

Introduction

The current interest in the cholesterol content of biological tissues and fluids stimulated a desire for better cholesterol methods (1, 2, 3). As a result, old techniques were modified (4, 5, 6, 7, 8, 9, 10) and new procedures proposed (11, 12, 13, 14, 15, 16) to satisfy the needs of routine and research laboratories (17, 18, 19, 20). The use of ferric chloride in a milieu of glacial acetic and sulfuric acids was suggested as a color reagent in a clinical method for determining serum cholesterol (21, 22, 23, 24, 25). Since the ferric chloride procedure was introduced (21), various modifications have appeared. When compared with classical procedures these were proved to be accurate (26, 27, 28, 29, 30, 31, 32, 33, 34, 35, 36, 37, 38).

Principles

Cholesterol can be extracted from serum with single or mixed organic solvents which remove it from its binding sites on the globulins. The solvent can be boiled off prior to reaction with ferric chloride solution in glacial acetic-sulfuric acid (22); or the reaction can take place in the extracting solvent (39). On reaction with the color reagent, a characteristic halochromic salt is produced wherein the purple chromophoric grouping is sensitive, stable, and reproducible (22, 23, 24). The direct determination of cholesterol in serum with-

* Based on the modifications of the method of Zlatkis, Zak, and Boyle (21).

out prior removal of protein is possible also (21, 40, 41, 42, 43), but error due to contamination of the reagent chemicals may present some difficulties (40).

Reagents

1. *Alcohol-acetone* (*protein precipitating agent*). Mix absolute alcohol and acetone in 1:1 proportions (33).

2. *Digitonin solution.* Add 1 g. of digitonin to 50 ml. of absolute ethanol and shake the mixture vigorously. Dilute to 100 ml. with distilled water (24). The addition of the water will dissolve the digitonin. This solution is stable for more than a month at room temperature.

3. *Cholesterol stock standard.* Dissolve 100 mg. of cholesterol per 100 ml. of high purity glacial acetic acid. See page 91 for the preparation of pure cholesterol. This solution is stable at room temperature.

4. *Ferric ammonium chloride reagent.* Dissolve 2.12 g. of $FeCl_3 \cdot 2NH_4Cl \cdot H_2O$ in 100 ml. of 80% (v/v) acetic acid (44).[1] This reagent is turbid because of slow solubility but clears on standing overnight. It is stable for months when stored in a brown bottle at room temperature.

5. *Ferric chloride precipitating agent.* Dissolve 850 mg. of $FeCl_3 \cdot 6H_2O$ in glacial acetic acid and dilute to 1 l. with glacial acetic acid. It is stable for several weeks when stored in a brown bottle at room temperature.

6. *Ferric chloride diluting agent.* Dilute 85 ml. of the ferric chloride precipitating agent to 100 ml. with glacial acetic acid. It is stable for several weeks when stored in a brown bottle at room temperature.

7. *Calibration standards* (*alcohol-acetone technique*). Dilute 8.00 ml. of the cholesterol stock standard to 10.00 ml. with glacial acetic acid. Prepare calibration standards containing 80, 160, and 240 μg. per 3 ml. in glacial acetic acid. These standards represent 100, 200, and 300 mg. of cholesterol per 100 ml. of serum in the procedure for total cholesterol, and 67, 133, and 200 mg. of cholesterol per 100 ml. of serum in the procedure for free cholesterol. Treat 3.00 ml. aliquots as described in the procedure beginning with the addition of 0.10

[1] The iron compound is available from several companies: Matheson, Coleman and Bell (Norwood, Ohio), Aceto Chemical Company (Flushing, N.J.), California Corporation for Biochemicals Research (Los Angeles, California), and K & K Laboratories (Plainview, Long Island, New York).

ml. of the ferric ammonium chloride solution. Prepare the calibration standards monthly.

8. *Calibration standards (ferric chloride-acetic acid precipitation technique).* Place 10.00 ml. of the cholesterol stock standard and 85.0 ml. of the ferric chloride precipitating agent into a 100 ml. volumetric flask and dilute to the mark with glacial acetic acid. Dilute 1.00, 2.00, 3.00 and 4.00 ml. to 4.00 ml. with ferric chloride diluting agent (reagent 6). These standards represent 100, 200, 300, and 400 mg. of cholesterol per 100 ml. of serum in the procedure. Pipet 3.00 ml. aliquots into 25 × 100 mm. test tubes and treat each as described for filtrate beginning with the layering in of 2.00 ml. of concentrated sulfuric acid. Prepare the calibration standards monthly.

NOTE: One Checker concluded that purification of the sulfuric acid by distillation over potassium dichromate was necessary (R. F. W.). Two others (W. E. C. and C. D. E.) found no significant difference in results obtained with reagent grade acetic further purified by redistillation and reagent grade (unpurified) acetic acid. Checker (M. C.) did not purify the reagent grade acetic acid but showed that there was no statistical difference between the proposed procedures and the Schoenheimer-Sperry procedure.

Procedures

TOTAL CHOLESTEROL (FERRIC CHLORIDE-ACETIC ACID PRECIPITATION TECHNIQUE

Pipet 0.10 ml. of serum into 4.00 ml. of the ferric chloride precipitating reagent. After 2–3 minutes, centrifuge the mixture for 10 minutes. Pipet 3.00 ml. of the filtrate into a 25 × 100 mm. test tube. Carefully layer in 2.00 ml. of concentrated sulfuric acid[2] and mix thoroughly by buzzing.

NOTE: Buzzing may be accomplished with a vortex mixer, or by a small Teflon-covered magnet and a magnetic stirrer.

When the solution has cooled to room temperature (approximately 20 minutes), measure the absorbance at 560 mμ in a 10 × 75 mm. (8 mm. I.D.) cuvet against a blank solution prepared from a mixture of 3.00 ml. of ferric chloride diluting agent and 2.00 ml. of concentrated sulfuric acid.[3] The color is stable for several hours and unaffected by ordinary light.

[2] Mallinckrodt A. R. sulfuric and acetic acids were found to be satisfactory.
[3] A Coleman Jr. spectrophotometer with an 8 mm. (I.D.) cuvet was used for all measurements described.

NOTE: The absorbances obtained here with an 8 mm. I.D. (10 × 75 mm.) cuvet were 0.0–1.0 for the 0–400 mg. per 100 ml. standards. The use of this range (0.25 absorbance per 100 mg. per 100 ml.) is based on the determination of a Ringbom plot (45). When more sensitive instruments or instruments with no adapters for cuvets with short light paths were used, the Checkers have found that the volume of precipitating agent should be increased. The standards are each diluted to 10 ml. instead of 4.0 ml., and the samples (0.1 ml.) are each diluted to 10 ml. with the precipitating agent in a volumetric flask.

NOTE: The presence of bubbles in the final chromophoric solution will alter the values of the absorbance to be read. These bubbles can be removed by centrifugation for 1–2 minutes.

TOTAL CHOLESTEROL (ALCOHOL-ACETONE TECHNIQUE)

Pipet 1.00 ml. of serum into approximately 10 ml. of alcohol-acetone extraction solution in a 25 ml. volumetric flask. Dilute the mixture to the mark with the solvents and shake it vigorously for approximately 1 minute. Filter through Whatman No. 41-H filter paper into a small-mouth bottle using a covered funnel to avoid evaporation, or centrifuge the precipitated proteins in a screw-capped tube for 10 minutes. Evaporate 2.00 ml. of the filtrate to dryness in a 25 × 100 mm. test tube. Use a boiling water bath as the source of heat. Pipet in 3.00 ml. of glacial acetic acid to dissolve the residue. Add exactly 0.10 ml. of the ferric ammonium chloride solution with mixing and then layer in 2.00 ml. of concentrated sulfuric acid. Mix the solution by buzzing, allow it to cool to room temperature, and measure its absorbance at 560 mμ against a blank containing the last three reagents with the volumes given.

NOTE: When less absorbance is necessary, a smaller aliquot (1 ml.) is evaporated to dryness. The working standards must be diluted 1 + 1 with glacial acetic acid to contain 40, 80, and 120 μg. per 3 ml. They still represent 100, 200, and 300 mg. per 100 ml. of serum.

FREE CHOLESTEROL (ALCOHOL-ACETONE TECHNIQUE)

Evaporate 3.00 ml. of the alcohol-acetone extract to approximately 0.5 ml. in a water bath. Add 1.0 ml. of the 1% digitonin solution, mix well, let the tube stand for 10–15 minutes, and centrifuge the digitonide for 10 minutes. Decant the supernatant fluid, and drain the inverted tube for several minutes. Blow in 4.0 ml. of acetone to disperse the precipitate. Centrifuge the washed digitonide and drain the tube until dry. Pipet in 3.00 ml. of glacial acetic acid and 0.10 ml.

of the ferric ammonium chloride solution and mix the solution before
layering in 2.00 ml. of concentrated sulfuric acid. Mix the solution
by buzzing until it is homogeneous. Then read the solution as previously described for total cholesterol.

NOTE: A smaller aliquot of the extract can be used for more sensitive absorbance measuring equipment.

Calculations

The cholesterol content can be determined from the absorbance-
concentration curves prepared with the described standards. Because the curves are linear, calculations can be made with one of the
standards from the particular technique used.

$$\text{Concentration}_{(unk)} \text{ in mg./100 ml.} = \frac{A_{unk}}{A_{std}} \times \text{concentration}_{(std)} \text{ in mg./100 ml.}$$

Discussion

CHOICE OF EXTRACTION TECHNIQUE

The original method for the determination of cholesterol with a
ferric chloride color reagent was direct (21). However, the possibility
of encountering a Hopkins-Cole reaction (46) with glyoxal impurities
in the acetic acid, and the tryptophan from protein, complicated the
procedure. Still, several investigators found the original method useful
when the acetic acid was purified by redistillation over potassium
dichromate (40, 41, 42, 43). A protein-free filtrate technique using
ferric chloride in glacial acetic acid as the precipitating agent was
investigated by several groups (47, 48, 49). Various organic solvents
or solvent mixtures have been used for the preparation of extracts or
protein-free filtrates of the sample. These include ethanol-acetone
(24), isopropanol (25), ethanol-ethyl acetate (39), chloroform-ethanol
(44), ethanol-ether (22), and ethanol (50). Several may be analyzed
directly without evaporation.

CHOICE OF COLOR REAGENT COMPOSITION

The original ferric chloride reagent was prepared by dissolving
ferric chloride in acetic acid and mixing an aliquot with concentrated sulfuric acid. At room temperature, this solution apparently lost

hydrochloric acid and formed some insoluble ferric sulfate which reduced the potency of the color reagent. Rosenthal (51) added phosphoric acid which maintained the constancy of the reagent by complexing the iron. Others found that the reagent could be stabilized by putting the iron in acetic acid (25), or by the use of ferric ammonium chloride in 80% acetic acid (44). In the first method for total cholesterol presented in this chapter, the color reagent also serves as the precipitating agent. The iron solution in 80% acetic acid (used for the analysis of total and free cholesterol) is quite stable. In addition, ferric ammonium chloride is preferable to ferric chloride because it is less hygroscopic and is therefore easier to weigh.

INTERFERENCES AND PRECAUTIONS

Bromide. Bromide enhances the color reaction in the ferric chloride color reaction (52). The approaches proposed for eliminating this interference are noteworthy. Rice and Lukasiewicz (52) removed bromide with silver iodate by metathetical exchange. The silver compound also eliminated interference due to thiouracil sometimes used to preserve serum for butanol-extractible iodine analysis (53). If cholesterol is determined on an aliquot of the serum before the thiouracil is added, this is an unnecessary precaution. Chiamori and Henry (49) removed the bromide by an ion exchange resin technique. Zak and Epstein (44) pretreated serum with ethanol, added alkali, and then extracted with chloroform to remove bromide and tryptophan. The partition coefficient favored the aqueous phase for the interfering substances. Bowman and Wolf (54) found that adding bromide to the glacial acetic acid enhanced the color reaction to a plateau so that bromide in the sample had no further effect. The same workers also showed that interference from bromide could be avoided by substituting ethanol for acetic acid in the color reaction (47).

Vitamin A. Vitamin A presumably interferes with the ferric chloride reaction (55, 56). A careful investigation of this constituent has shown that these early reports were in error. No significant interference can be shown when vitamin A is added in excess to serum prior to carrying out the determination (40, 44).

NOTE: Checker (M. C.) found no interference caused by vitamin A.

Bilirubin. Bilirubin is only partially extracted by several of the existing solvent systems. It remains in the aqueous phase when extraction is carried out from alkaline solution in which bilirubin is more soluble. Bilirubin may be removed when aluminum hydroxide is used as an adsorbent in the extraction reagent (57). When part of the bilirubin is extracted into the organic phase in any particular system used, it is oxidized to biliverdin by the subsequent ferric chloride reaction (58). The spectral characteristics of biliverdin show a minimum where cholesterol shows a maximum (21), and a much lower absorptivity.

Tryptophan. The tryptophan of intact protein can give a Hopkins-Cole reaction if glyoxal is present in the reaction system (21). The amount of free tryptophan in serum is too low to seriously affect the determination. This protein-tryptophan reaction can be avoided by purification of the acetic acid used in the procedure (21, 40, 42, 45).

Nitrate. Nitrate supposedly interferes with the ferric chloride reaction (3). However, no information is available on why this happens, nor has this finding been confirmed.

SAPONIFICATION

Saponification of cholesterol esters is a common step in several procedures (33, 34, 35). A number of papers indicate that saponification is unnecessary for routine analysis of cholesterol by a ferric chloride modification (24, 26, 28, 29, 31). However, there are pro and con data (18, 22, 27, 59, 60, 61, 62) on whether cholesterol and its esters react to give the same amount of color per unit of cholesterol. Most of the studies cited here indicate that free and esterified cholesterol yield equal color densities. There is a modification which indicates that free and esterified cholesterol can be made to react at different rates by temperature control. Girard and Assous (30, 63) found that carrying out the color reaction at 20°C. gave complete halochromic salt formation with only free cholesterol. Elevation of the temperature to 65°C. gave the proper chromophoric equivalent for the esters.

Normal Ranges

The normal range determined by the use of the described procedures was 220 ±50 mg. per 100 ml. of serum for all adults.

NOTE: This is in good accord with the over-all values found by Checker (C. D. E.), which are given in the tabulation.

Age	No. of samples	Total A[a] Mean ± S.D.	Total B[b] Mean ± S.D.	Free[b] Mean ± S.D.
19–29	29	200 ± 37	193 ± 37	52 ± 11
30–39	28	226 ± 43	230 ± 40	61 ± 15
40–49	25	244 ± 48	236 ± 41	56 ± 12
50–59	15	238 ± 52	233 ± 57	66 ± 19
60–76	7	222 ± 35	199 ± 21	57 ± 9

[a] Ferric chloride-acetic acid protein precipitation technique.
[b] Alcohol-acetone protein precipitation technique.

The average normal value for 105 samples was 225 ± 47 mg./100 ml. for the total A technique and 219 ± 45 mg./100 ml. for the total B technique on 103 samples. Free cholesterol on the same 103 samples was 57 ± 14.

REFERENCES

1. Kabara, J. J., Determination and microscopic localization of cholesterol. *Methods Biochem. Anal.* **10**, 263–318 (1962).
2. Cook, R. P., "Cholesterol." Academic Press, New York, 1958.
3. Kritchevsky, D., "Cholesterol." Wiley, New York, 1958.
4. Cramer, K., and Isaksson, B., An evaluation of the Theorell method for the determination of total serum cholesterol. *Scand. J. Clin. Lab. Invest.* **11**, 213–216 (1959).
5. Kenny, A. P., The determination of cholesterol by The Liebermann-Burchard reaction. *Biochem. J.* **52**, 611–619 (1952).
6. Saifer, A., Photometric determination of total and free cholesterol and the cholesterol ester ratio of serum by a modified Liebermann-Burchard reaction. *Am. J. Clin. Pathol.* **21**, 24–32 (1951).
7. Martensson, E. H., Investigation of factors affecting the Liebermann-Burchard cholesterol reaction. *Scand. J. Clin. Lab. Invest.* **15**, *Suppl. 69*, 164–180 (1963).
8. Ferro, P. V., and Ham, A. B., Rapid determination of total and free cholesterol in serum. *Am. J. Clin. Pathol.* **33**, 545–549 (1960).
9. Jørgensen, K. H., and Dam, H. An ultramicromethod for the determination of total cholesterol in bile based on the Tschugaeff color reaction. *Acta Chem. Scand.* **11**, 1201–1208 (1958).
10. Sobel, A. E., Goodman, J., and Blaw, M., Cholesterol in blood serum. Studies of microestimation as the pyridinium cholesteryl sulfate. *Anal. Chem.* **23**, 516–519 (1951).
11. Brown, W. D., Determination of serum cholesterol with perchloric acid. *J. Exptl. Biol.* **37**, 523–532 (1959).
12. Weigensberg, B. I., and McMillan, G. C., Ultraviolet spectrophotometric method for the determination of cholesterol. *Am. J. Clin. Pathol.* **31**, 16–25 (1959).
13. Pearson, S., Stern, S., and McGavack, T. H., A rapid procedure for the

determination of serum cholesterol. *J. Clin. Endocrinol. Metabolism* **12**, 1245–1246 (1952).

14. Jurand, J., and Albert-Recht, F., The estimation of serum cholesterol. *Clin. Chim. Acta* **7**, 522–528 (1962).

15. Pollak, O. J., and Wadler, B., Rapid turbidimetric assay of cholesterols. *J. Lab. Clin. Med.* **39**, 791–794 (1952).

16. Vahouny, G. V., Borja, C. R., Mayer, R. M., and Treadwell, C. R., A rapid quantitative determination of total and free cholesterol with anthrone reagent. *Anal. Biochem.* **1**, 371–581 (1960).

17. Burstein, M., and Samaille, J., Sur un dosage rapide du cholesterol lie aux α- et aux β-lipoprotéines du sérum. *Clin. Chim. Acta* **5**, 609 (1960).

18. Hansen, P. W., and Dam, H., Paper chromatography and colorimetric determination of free and esterified cholesterol in very small amounts of blood. *Acta Chem. Scand.* **11**, 1658–1662 (1957).

19. Abell, L. L., and Mosbach, E. H., Hypercholesterolemic effect of methytestosterone in rats. *J. Lipid Res.* **3**, 88–91 (1962).

20. Swartout, J. R., Dieckert, J. W., Miller, D. N., and Hamilton, J. G., Quantitative glass paper chromatography: A microdetermination of plasma cholesterol. *J. Lipid Res.* **1**, 281–285 (1960).

21. Zlatkis, A., Zak, B., and Boyle, A. J., A new method for the direct determination of serum cholesterol. *J. Lab. Clin. Med.* **41**, 486–492 (1953).

22. Crawford, N., An improved method for the determination of free and total cholesterol using the ferric chloride reaction. *Clin. Chim. Acta* **3**, 357–367 (1958).

23. Yoneyama, T., Kitamura, M., and Yoshikawa, H., Normal values of total serum cholesterol in healthy Tokyo citizens. *Clin. Chim. Acta* **7**, 529–536 (1962).

24. Zak, B., Dickenman, R. C., White, E. G., Burnett, H., and Cherney, P. J., Rapid estimation of free and total cholesterol. *Am. J. Clin. Pathol.* **24**, 1307–1315 (1954).

25. Leffler, H. H., Estimation of cholesterol in serum. *Am. J. Clin. Pathol.* **31**, 310–313 (1959).

26. Smit, Z. M., and Wehmeyer, A. S., A comparative study of two methods for the determination of serum cholesterol. *S. African J. Lab. Clin. Med.* **3**, 321–325 (1957).

27. Webster, D., The determination of total and ester cholesterol in whole blood, serum or plasma. *Clin. Chim. Acta* **1**, 277–284 (1962).

28. Wilkens, J. A., and De Wit, H., A micromethod for the determination of cholesterol in 0.2 ml. of blood. *S. African J. Lab. Clin. Med.* **7**, 13–16 (1961).

29. Langan, T. A., Durrum, E. L., and Jencks, W. P., Paper electrophoresis as a quantitative method: Measurement of alpha and beta lipoprotein cholesterol. *J. Clin. Invest.* **34**, 1427–1436 (1955).

30. Girard, M., and Assous, E., Possibilité d' évaluer, sans séparation préalable, les fractions libre et esterifieé du cholestérol total. *Bull. Soc. Chim. Biol.* **43**, 1097–1109 (1961).

31. Courchains, A. J., Miller, W. H., and Stein, Jr., D. B., Rapid semimicro procedure for estimating free and total cholesterol. *Clin. Chem.* **5**, 609–614 (1959).

32. Jérome, H., Fonty, P., Vernin, H., and Leplaideur, F., Étude d'une méthode simplifiée de dosage du cholestérol total et libre dans le sérum sanguin par la réaction au chlorure ferrique. *Ann. Biol. Clin.* (*Paris*) **20**, 117–133 (1962).
33. Schoenheimer, R., and Sperry, W. M., A micromethod for the determination of free and combined cholesterol. *J. Biol. Chem.* **106**, 745–760 (1934).
34. Sperry, W. M., and Webb, M., A revision of the Schoenheimer and Sperry method for cholesterol determination. *J. Biol. Chem.* **187**, 97–106 (1950).
35. Abell, L. L., Levy, B. B., Brodie, B. B., and Kendall, F. E., Simplified method for the estimation of total cholesterol in serum and demonstration of its specificity. *J. Biol. Chem.* **195**, 357–366 (1952).
36. Sackett, G. E., Modification of Bloor's method for the determination of cholesterol in whole blood or blood serum. *J. Biol. Chem.* **64**, 203–205 (1925).
37. Bloor, W. R., The determination of cholesterol in blood. *J. Biol. Chem.* **24**, 227–231 (1916).
38. Paget, M., and Pierrart, G., Recherches sur le dosage du cholestérol et de ses esters. *Bull. Soc. Chim. Biol.* **21**, 537–548 (1939).
39. Klungsöyr, L., Hawkenes, E., and Kloss, K., A method for the determination of cholesterol in blood serum. *Clin. Chim. Acta* **3**, 514–518 (1958).
40. Moore, R. V., and Boyle, Jr., E., Errors associated with the direct measurement of human serum cholesterol using the $FeCl_3$ reagent. *Clin. Chim. Acta* **9**, 156–162 (1963).
41. Figueroa, I. S., and Pike, R. L., An ultramicromethod for the determination of total serum cholesterol. *Anal. Biochem.* **2**, 103–106 (1960).
42. MacIntyre, I., and Ralston, M., Direct determination of serum cholesterol. *Biochem. J.* **56**, xliii (1954).
43. Debot, M. P., Sur le dosage du cholestérol serique. *J. Med. Bordeaux Sud-Ouest* **137**, 253–259 (1960).
44. Zak, B., and Epstein, E., A study of several color reactions for the determination of cholesterol. *Clin. Chim. Acta* **6**, 72–78 (1961).
45. Willard, H. H., Merritt, L. L., Jr., and Dean, J. A., "Instrumental Methods of Analysis," 3rd ed., pp. 47–50. Van Nostrand, New York, 1958.
46. Hopkins, F. E., and Cole, S. W., On the proteid reaction of Adamkiewicz with contributions to the chemistry of glyoxylic acid. *Proc. Roy. Soc.* (*London*) **68**, 21–23 (1901).
47. Zak, B., Simple rapid microtechnic for serum total cholesterol. *Am. J. Clin. Pathol.* **27**, 583–588 (1957).
48. Henley, A. A., The determination of serum cholesterol. *Analyst* **82**, 286–287 (1957).
49. Chiamori, N., and Henry, R. J., Study of the ferric chloride method for the determination of total cholesterol and cholesterol esters. *Am. J. Clin. Pathol.* **31**, 305–309.
50. Bowman, R. E., and Wolf, R. C., A rapid and specific ultramicromethod for total serum cholesterol. *Clin. Chem.* **8**, 302–309 (1962).
51. Rosenthal, H. L., Pfluke, M. L., and Buscaglia, S., A stable iron reagent for determination of cholesterol. *J. Lab. Clin. Med.* **50**, 318–321 (1957).
52. Rice, E. W., and Lukasiewicz, D. B., Interference of bromide in the Zak

ferric chloride-sulfuric acid cholesterol method, and means of eliminating this interference. *Clin. Chem.* 3, 160–162 (1957).

53. Rice, E. W., Interference of thiouracil in the Zak cholesterol method and elimination of this interference with silver iodate. *Clin. Chem.* 9, 493 (1963).

54. Bowman, R. E., and Wolf, R. C., Observations on the Zak cholesterol reaction: Glacial acetic acid purity. a more sensitive absorbance peak, and bromide enhancement. *Clin. Chem.* 8, 296–301 (1962).

55. Kinley, L. J., and Krause, R. F., Serum cholesterol determinations as affected by vitamin A. *Proc. Soc. Exptl. Biol. Med.* 99, 244–245 (1958).

56. Kinley, L. J., and Krause, R. F., Influence of vitamin A on cholesterol blood levels. *Proc. Soc. Exptl. Biol. Med.* 102, 353–355 (1959).

57. Babson, A. L., Shapiro, P. O., and Phillips, G. E., A new assay for cholesterol and cholesterol esters in serum which is not affected by bilirubin. *Clin. Chim. Acta* 7, 800–804 (1962).

58. Zak, B., Moss, N., Boyle, A. J., and Zlatkis, A., Spectrophotometric study of some oxidative products of bilirubin. *Anal. Chem.* 26, 1220–1222 (1954).

59. Shin, R. S., Silicic acid column chromatography for the microdetermination of cholesterol, cholesterol ester, and phospholipid from human cerebrospinal fluid. *Anal. Biochem.* 5, 369–378 (1963).

60. Wycoff, H. D., and Parson, J., Chromatographic microassay for cholesterol and cholesterol esters. *Science* 125, 347 (1957).

61. Kurzweg, G., and Massmann, W., About the differences of extinction of the free and esterified cholesterol using a few simple methods of estimation. *Clin. Chim. Acta* 7, 515–521 (1962).

62. Cohen, L., Jones, R. J., and Batra, K. V., Determination of total, ester and free cholesterol in serum and serum lipoproteins. *Clin. Chim. Acta* 6, 613–619 (1961).

63. Assous, E. F., and Girard, M. L., Microméthode nouvelle de dosage du cholestérol libre et du cholestérol total directement sur le sérum sanguin. *Ann. Biol. Clin. (Paris)* 20, 973–986 (1962).

CHOLESTEROL (PRIMARY STANDARD)

Submitted by: NATHAN RADIN, The Rochester General Hospital, Rochester, New York

Checked by: OLIER L. BARIL, College of the Holy Cross, Worcester Massachusetts

MARSHALL E. DEUTSCH, Diagnostic Research Center, Downingtown, Pennsylvania

Introduction

Primary standards are chemical substances which, by virtue of their purity, can be weighed directly for the preparation of solutions with known concentrations. Purity of chemical compounds may be defined in various ways. Ideally, a pure compound is one in which all molecules are identical. For the preparation of cholesterol, however, purity is defined in the practical terms of the purification steps performed. A system of molecules is considered a pure compound if an exhaustive series of fractionations fails to produce fractions with different properties. The criteria used to define a pure compound thus change as new methods become available for fractionating material, or for more accurately measuring the properties of the fractions so produced (1).

Commercial cholesterol is usually prepared by solvent extraction of the spinal cords and brains of cattle. Cholesterol is then crystallized and recrystallized directly or after the esterified cholesterol has been saponified with alkali. Cholesterol, as usually prepared, is contaminated by cholestanol (dihydrocholesterol or cholestan-3β-ol), cholest-7-en-3β-ol (lathosterol or Δ^7-cholestenol), and traces of 7-dehydrocholesterol (cholesta-5,7-dien-3β-ol) (2).

Procedures are presented for the recrystallization of cholesterol from ethanol, the recrystallization of cholesterol from acetic acid (3), and the treatment of cholesterol in ether solution with bromine, precipitating the insoluble 5,6β-dibromide derivative, which is then regenerated with zinc and acetic acid (4). Methods for the characterization of cholesterol by the melting point (5), the molar absorptivity with the Liebermann-Burchard reaction (6), and the molar absorptivity with the sulfuric acid-iron reaction (7) will also be described.

91

Reagents

Cholesterol. Any commercial product with at least U.S.P. specifications (8).

For cholesterol recrystallization

From ethanol:
1. *Absolute ethanol.*
2. *Diethyl ether.*
From acetic acid:
1. *Glacial acetic acid.*
2. *Methanol.*
With the dibromide method:
1. *Glacial acetic acid.*
2. *Diethyl ether.*
3. *Brominating solution.* Dissolve 22.7 g. (7.75 ml.) of bromine and 1.67 g. of anhydrous sodium acetate in 200 ml. of glacial acetic acid.
4. *Zinc dust.*
5. *Acid solution.* Mix 18.0 ml. of concentrated hydrochloric acid with 300 ml. of water.
6. *Sodium hydroxide solution, 10%.* Dissolve 50.0 g. of sodium hydroxide pellets in 500 ml. of water.
7. *Methanol.*

For characterization of cholesterol by the Liebermann-Burchard reaction

1. *Glacial acetic acid.*
2. *Acetic anhydride.*
3. *Sulfuric acid-glacial acetic acid solution, 20% (v/v).* Slowly add 100 ml. of concentrated sulfuric acid to about 100 ml. of glacial acetic acid in a 500 ml. volumetric flask. Keep the flask in cool water while adding the sulfuric acid in order to prevent excessive heating during the mixing operation. When all of the sulfuric acid has been added and the contents of the flask are at room temperature dilute to the mark with glacial acetic acid.

For characterization of cholesterol by the sulfuric acid-iron reaction:

1. *Glacial acetic acid.*
2. *Ferric chloride stock solution.* Dissolve and dilute 5 g. of ferric

chloride, $FeCl_3 \cdot 6H_2O$, with concentrated orthophosphoric acid (85%) to the mark of a 50 ml. volumetric flask.

3. *Ferric chloride color reagent solution.* Dilute 5.0 ml. of the ferric chloride stock solution and 15.0 ml. of concentrated phosphoric acid with concentrated sulfuric acid to the mark of a 250 ml. volumetric flask. Prepare this dilute color reagent freshly on the day of use.

NOTE: All of the chemicals described should meet the specifications of the American Chemical Society (9).

Apparatus for Characterization of Cholesterol

Use the Beckman spectrophotometer (Model DU, with matched absorption cells, 10 mm. light path). Adjust the slit width after the sensitivity knob has been rotated 5 full turns from the extreme clockwise position. Keep the cell compartment at $25 \pm 0.5°C.$, if possible, as the molar absoptivity values are temperature dependent.

Cholesterol Recrystallization Methods

CHOLESTEROL RECRYSTALLIZATION FROM ETHANOL

Add cholesterol in the proportion of 1 g./ml. to absolute ethanol. Heat a beaker containing the mixture on a hot plate, with mixing by swirling, until the cholesterol is dissolved. Then remove the beaker from the hot plate, cover it with a watch glass, and allow it to cool to room temperature. Separate the precipitated cholesterol by filtering in a Büchner funnel through Whatman No. 1 filter paper. Wash the damp crystals with a small volume of diethyl ether, dry overnight in air, then dry in an oven at 90°C. for 2 hours. Store in a brown bottle under refrigeration (4–10°C.).

CHOLESTEROL RECRYSTALLIZATION FROM ACETIC ACID

Heat glacial acetic acid to the boiling point in an Erlenmeyer flask and then pour it into a beaker containing cholesterol. Use 8.0 ml. of acetic acid for each gram of cholesterol. Stir the mixture vigorously with a wooden paddle until the solid material is all dissolved. Then plunge the beaker into an ice bath, while stirring the contents vigorously. When room temperature is attained collect the complex on Whatman No. 1 filter paper in a Büchner funnel. After washing with acetic acid and methanol, air-dry the crystals overnight, then dry in an oven at 90°C. for 2 hours. Store the cholesterol in a brown bottle under refrigeration (4–10°C.).

RECRYSTALLIZATION OF CHOLESTEROL WITH THE DIBROMIDE METHOD

Dissolve 50 g. of cholesterol in 350 ml. of diethyl ether by gentle warming on a steam bath. After cooling the resultant solution to room temperature add 200 ml. of the brominating solution. A white paste results. Stir the mixture about 10 minutes while cooling to 20°C. in an ice bath. Transfer the material to a Büchner funnel containing Whatman No. 1 filter paper and wash it with acetic acid until the washings are colorless. Suspend the white material in 750 ml. of diethyl ether and 10 ml. of glacial acetic acid. Slowly add 10 g. of zinc dust. A reaction is evidenced by bubbles of hydrogen. If necessary, warm cautiously on a steam bath. Continue until the evolution of hydrogen is complete (half an hour or more). A white precipitate of zinc salts forms. Add 10 ml. of water to dissolve the precipitate. Decant the ether solution from the water layer containing the excess solid zinc. Wash the ether solution twice in a separatory funnel with 150-ml. portions of acid solution and then three times with 150-ml. portions of the sodium hydroxide solution to neutralize the acetic acid. Add 500 ml. of methanol to the ether solution and then evaporate the solution slowly on a steam bath until most of the ether is removed and the purified cholesterol begins to crystallize. The crystallization then proceeds at room temperature. Collect the product on Whatman No. 1 filter paper in a Büchner funnel, air-dry it overnight, and then dry it in an oven at 90°C. for 2 hours. Store the material in a brown bottle under refrigeration (4–10°C.).

NOTE: Anhydrous cholesterol, obtained by crystallization from dry solvents, forms triclinic needles. Cholesterol separates as monohydric, rhombic, triclinic plates from moist solvents (2). Ethanol-recrystallized cholesterol forms a white, fluffy crystalline material. Acetic acid-recrystallized cholesterol forms a white, packed material. Cholesterol obtained from dibromide recrystallization appears as a lustrous white powder.

The cholesterol monohydrate loses water at 70–80°C. (2). In order to ensure anhydrous cholesterol after air-drying, the recrystallized preparations are heated in an oven at 90°C. for 2 hours. Cholesterol preparations treated this way show no change from the original appearance. However. it is better to use a vacuum oven for this operation.

Weight changes due to the loss of moisture are negligible after commercial cholesterol preparations are heated and dried.

NOTE: Checker (O. L. B.) reports that the crystallization methods given here are not applicable to the purification of as little as 1 g. of cholesterol. Checker (M. E. D.), however, considers 5 g. feasible. The purification of 25 g. or 50 g. of cholesterol at one time by any of the methods is recommended.

Cholesterol Characterization Methods

MELTING POINT

The method for obtaining melting points in evacuated capillaries is described by Fieser (5). Fieser has observed that the melting points taken in open capillaries are often several degrees lower than those taken in evacuated capillaries.

Push the open end of a melting point capillary into the powdered cholesterol and tamp down the solid until there is a column about 5 mm. long. Connect the tube to a suction pump with an adapter consisting of a glass tube capped with a one-holed rubber vaccine stopper. Make the hole with an awl, ice pick, or red-hot hypodermic needle. Lubricate the end of the hole extending toward the wider part of the stopper with water, grasp the capillary very close to the sealed end, and force it through nearly to the full length. Connect it to the suction pump and make a guiding mark 1.5 cm. from the closed end where a seal is to be formed. Grasp the end of the tube with one hand and the adapter with the other, holding the point to be sealed near the small flame of a microburner. Steady both hands on the bench so that neither hand will move much when the tube is melted. Hold the tube in the flame until the walls collapse to form a flat seal and then remove the tube at once. Take the tube out of the adapter and, if the two straight parts are not in a line, soften the seal and correct the alignment. Attach the capillary to a calibrated thermometer with a rubber band. Insert this assembly into a Thiele tube which contains glycerol or mineral oil. Heat the Thiele tube rapidly with a microburner until the temperature is about 125°C., then reduce to a rate of about 2°C. per minute. Observe both the sample and the thermometer and record the temperatures of initial and terminal melting.

Fieser has shown that the melting point for cholesterol purified by recrystallization from acetic acid is 149.5–150°C. (3). The data shown by Radin and Gramza (10) indicate that for recrystallized cholesterol the acceptable melting point range is 149.3–151.3°C.

LIEBERMANN-BURCHARD REACTION

A modified Liebermann-Burchard procedure was adapted from the Carr and Drekter (6) cholesterol technique.

Use the following procedure to measure the absorbance and to calculate the molar absorptivity:

1. Transfer about a 0.1000 g. portion of the cholesterol preparation to a 100 ml. volumetric flask using glacial acetic acid as the wash solvent. Keep the flask at room temperature overnight to permit the complete solution of the cholesterol. Then dilute the solution to mark with glacial acetic acid and mix well.

2. Pipet a 5.00 ml. aliquot of the cholesterol solution into a 50 ml. volumetric flask. Pipet 5.00 ml. of glacial acetic acid into another 50 ml. volumetric flask for the reagent blank.

3. Pipet a 20.0 ml. aliquot of acetic anhydride (*Caution!*) into each volumetric flask.

4. Pour the 20% sulfuric acid-glacial acetic solution to about the neck of the reagent blank solution flask. After swirling to mix, place the flask in a water bath at 25°C. Pour the 20% sulfuric acid-glacial acetic acid solution up to about the neck of the flask containing the cholesterol solution while starting a stop watch. After swirling to mix, place the flask in a water bath at 25°C. Follow the same procedure with other cholesterol solutions at 1 minute intervals. Dilute each solution to mark with the acid mixture, mix, and return to the water bath at exactly 10 minutes after the initial operation.

5. Measure the absorbance of each solution at 620 mμ exactly 20 minutes after adding the 20% sulfuric acid-glacial acetic acid solution.

6. Calculate the molar absorptivity for each solution measured using the Bouguer-Beer law, $A = abc$, where A is absorbance; a, molar absorptivity; b, light path in centimeters; and c, concentration in moles per liter. The equation derived for the conditions described here is:

$$\text{Molar absorptivity} = 386.6 \times \frac{\text{absorbance}}{\text{cholesterol sample weight (g.)}}$$

7. The mean molar absorptivity, recommended as a guide to ascertain purity with the Liebermann-Burchard procedure, is 1700 ± 30 (1 S.D.) liters cm.$^{-1}$ moles^{-1}, at 25°C. (10).

SULFURIC ACID-IRON REACTION

The following procedure for the sulfuric acid-iron reaction was adapted from the Rosenthal, Pfluke, and Buscaglia (7) cholesterol technique.

1. Transfer about 0.1000 g. of the cholesterol preparation to a

100 ml. volumetric flask using glacial acetic acid as the wash solvent. Keep the flask at room temperature overnight to permit the complete solution of the cholesterol. Then dilute the solution to the mark with glacial acetic acid.

2. Dilute a 3.00 ml. aliquot of the cholesterol stock solution (Step 1) with glacial acetic acid to mark in a 100 ml. volumetric flask.

3. Transfer a 5.00 ml. aliquot of the dilute cholesterol solution (Step 2) into a 50 ml. Erlenmeyer flask. Pipet 5.00 ml. of glacial acetic acid into a 50 ml. Erlenmeyer flask for the blank. Place the flasks in a water bath at 25°C. Transfer 4.00 ml. aliquots of the dilute ferric chloride color reagent into each flask at 1 minute intervals and mix by swirling.

4. Measure the absorbance at 560 mμ 30 minutes after adding the dilute color reagent.

NOTE: Checker (M. E. D.) recommends that the color reagent be added in Step 3 at 2-minute intervals to allow sufficient time for mixing the reagents and the rinsing, and for filling and reading of the absorption cells. Any time interval is suitable provided the absorbance is measured 30 minutes after adding the color reagent.

5. Calculate the molar absorptivity using the Bouguer-Beer law, $A = abc$, where A is absorbance; a, molar absorptivity; b, light path in centimeters; and c, concentration in moles per liter. In this instance the concentration value is based on a 9 ml. final volume and is not corrected for any volume change that takes place when the glacial acetic acid and sulfuric acid solutions are mixed. The equation derived for the conditions described here is:

$$\text{Molar absorptivity} = 2320 \times \frac{\text{absorbance}}{\text{cholesterol sample weight (g.)}}$$

6. The mean molar absorptivity, recommended as a guide to ascertain purity with the sulfuric acid-iron reaction, is $11{,}500 \pm 100$ (1 S.D.) liters cm.$^{-1}$ moles^{-1}, at 25°C. (10).

Discussion

The melting points of unrecrystallized commercially obtained cholesterol preparations are generally lower than the melting points of the same materials after recrystallization. Recrystallized products have narrower melting point ranges than the original commercial products. However, the greater the melting point range for the original commercial product the greater the melting point range for the recrystallized product (10). This indicates that some of the impurities in the original commercial cholesterol preparations are not removed by recrystallization.

Information from various commercial sources is that the U.S.P. specifications (8) are now the accepted standard for cholesterol purity. The systematic investigation of recrystallization and characterization of cholesterol indicates that further studies of this kind are necessary (10) in order to develop better purification and characterization techniques than are now being used. However, for the present, three recrystallization methods are presented to enable the clinical chemist to prepare an adequate cholesterol standard. The dibromide derivative method is the method of choice because the cholesterol preparations obtained with it have the greatest molar absorptivity (10). According to Fieser (11), crystallization of cholesterol from ethanol does not remove the cholest-7-en-3β-ol and cholestanol present. Purification through the dibromide method, however, eliminates these contaminants as well as two products of air oxidation, 7-ketocholesterol and cholestane-3β,25-diol. The acetic acid recrystallization method for cholesterol is recommended in preference to alcohol method if the dibromide method cannot be carried out in the laboratory.

NOTE: The melting points and the molar absorptivities of five commercial cholesterol preparations did not meet the specifications recommended here (10). On the basis of the finding and other (unpublished) data, it is recommended that the purity of commercial cholesterol preparations should be checked. The preparations should be recrystallized if the melting point or the molar absorptivity does not meet the specifications described here.

REFERENCES

1. Eyring, H., Philosophy of the purity and identity of organic compounds. Anal. Chem. 20, 98–100 (1948).
2. Cook, R. P., "Cholesterol." Academic Press, New York, 1958.
3. Fieser, L. F., Cholesterol and companions. III. Cholestanol, lathosterol and ketone 104. J. Am. Chem. Soc. 75, 4395–4403 (1953).
4. Fieser, L. F., Cholesterol and companions. VII. Steroid dibromides. J. Am. Chem. Soc. 75, 5421–5422 (1953).
5. Fieser, L. F., "Experiments in Organic Chemistry," p. 23. Heath, Boston, Massachusetts, 1955.
6. Carr, J. J. and Drekter, I. J., Simplified rapid technique for the extraction and determination of serum cholesterol without saponification. Clin. Chem. 2, 353–368 (1956).

7. Rosenthal, J. L., Pfluke, M. L., and Buscaglia, S., A stable iron reagent for determination of cholesterol. *J. Lab. Clin. Med.* **50**, 318–322 (1957).
8. "The Pharmacopeia of the United States of America," 16th ed. The United States Pharmacopeial Convention, Inc., Washington, D.C., 1960.
9. American Chemical Society, Committee on Analytical Reagents, "Reagent Chemicals: The American Chemical Society Specifications," Washington, D.C., 1955.
10. Radin, N. and Gramza, A. L., Standard of purity for cholesterol. *Clin. Chem.* **9**, 121–134 (1963).
11. Fieser, L. F., Private communication.

CHLORIDE IN SWEAT*

Submitted by: FRANK A. IBBOTT, University of Colorado Medical Center, Denver, Colorado
Checked by: WALTER R. C. GOLDEN, The Stamford Hospital, Stamford, Connecticut
EMANUEL EPSTEIN, St. Joseph Mercy Hospital, Pontiac, Michigan
MATTHEW J. REHAK, St. Agnes Hospital, Baltimore, Maryland
JEROME M. WHITE, Santa Clara County Hospital, San Jose, California

Introduction

After the discovery that the sweat of patients with cystic fibrosis contains a high concentration of sodium and chloride (1, 2, 3), the determination of sweat chloride has become clinically important. The procedure may be divided into two main parts: sweat collection, and analysis of the sweat for chloride content. In order to obtain the minimum quantity of sweat required for a reliable chloride assay in a conveniently short time, it is necessary to deliberately induce sweating. One of the earlier methods was to place the patient for 30 to 90 minutes in a plastic suit with an elastic neck and cover him with a blanket (4). This thermal stimulus has also been supplemented by the use of hot water bottles (5) or electric lights (6). Objections to these methods have been raised due to the occurrence of hyperpyrexia in some patients (7) and to a fatal outcome in others (5, 8).

NOTE: Heat induction of sweating in one limb is permissible in children, including the newborn, provided that the patient is carefully observed for adverse signs (R. S. G., ed.).

Samples of sweat have been obtained without the application of external heat (9) or by wrapping the patient in a blanket (7).

Two parasympathomimetic agents, Mecholyl® and pilocarpine, have been used to produce sweating in localized areas. The intradermal injection of 2 mg. of Mecholyl into the ventral surface of the arm was

* Based on the method of Gibson and Cooke (11).

101

found to stimulate the production of sufficient sweat for sodium and chloride analysis (10). However, pilocarpine has been more extensively used for this purpose, particularly when coupled with iontophoresis, the process of introducing a drug through the intact skin by the application of a direct electric current (11). This procedure requires only a short time to complete, usually produces an adequate volume of sample, and can be used on patients of all ages.

For situations where an accurate assay of sweat chloride is not available, a simple semiquantitative procedure may give information of value. However, it is advisable to confirm such results by the quantitative method. The screening test consists of applying the palm of the patient's hand to a layer of an agar-silver chromate reagent. When the chloride concentration in the sweat is greater than approximately 60 meq./l., a color change is produced in the reagent (12, 13).

Apparatus

POWER SUPPLY

Suitable power sources are available commercially,[1,2] although it is not difficult to build one for very little cost. The parts required are listed below and the circuit diagram is given in Fig. 1.

1. *Metal utility cabinet, 6 × 5 × 4 in.*
2. *Three 7.5-volt batteries (e.g., Eveready No. 707).*
3. *Rheostat, 2500 ohms resistance.*
4. *Milliammeter, 5 ma. full-scale.*
5. *Switch.*
6. *Indicator light assembly.*
7. *Two electrodes.* These may be electrocardiograph electrodes or specially fabricated from stainless steel. Modified electrocardiograph electrodes[1] and a pattern incorporating an electrolyte reservoir[2] are available commercially. The leads connecting the electrodes to the power supply should be marked to indicate polarity; red for the positive lead, and black for the negative lead. When necessary, polish the positive electrode with a fine grade sandpaper to remove tarnish.

NOTE: It was found that, of the two pairs of electrodes supplied with the instrument made by the Buchler Instrument Company, only the 20 mm. diameter

[1] Buchler Instrument Company, Fort Lee, New Jersey.
[2] Heat Technology Laboratories, Inc., Alloyd Laboratory Branch, 4308 Governor's Drive, Huntsville, Alabama.

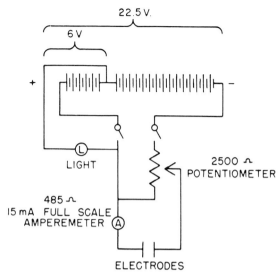

FIG. 1. Circuit diagram for iontophoresis power supply. (Reproduced from L. E. Gibson and R. E. Cooke (11) with permission.

pair was satisfactory. The larger pair produced burns with a current of 4 ma. (W. R. C. G.).

Materials

1. Plastic containers with covers. A plastic container with a friction-fit cover, 1-⅞ in. in diameter × 3-⅞ in. high, is convenient.[3]

NOTE: Glass containers with standard taper ground-glass stoppers are equally satisfactory (W. R. C. G.).

2. Filter paper circles, 2.5 cm. in diameter, Whatman No. 540 or No. 42.

NOTE: Schleicher and Schuell #589 Green Ribbon filter paper is suitable (E. E.). A single thickness of Johnson & Johnson "Topper" gauze may be substituted for the filter paper during iontophoresis (W. R. C. G.).

3. Pilocarpine nitrate, 0.2% (w/v). Dissolve 0.2 g. of pilocarpine nitrate in water and dilute to 100 ml. Store under refrigeration. The solution is stable for at least 3 months.

[3] The Nalge Company, Inc., Rochester 2, New York.

4. *Sodium chloride, 0.9% (w/v)*. Dissolve 0.9 g. NaCl in water and dilute to 100 ml.

5. *Deionized water.*

6. *Rubber straps or bands to secure the electrodes.* These may be the type supplied with electrocardiograph electrodes, or rubber bands 3 × ⅛ in.

7. *Masking tape or water-proof adhesive tape.*

8. *Five-cm. squares of Parafilm.*[4] Leave the protective paper layer in place.

9. *Paper tissues.*[5]

NOTE: The listed items are best maintained as a kit.

Preparation of Weighing Bottles

Immediately prior to a test remove the cap from a weighing bottle, handling it with tissues or a chamois cloth. Using forceps, insert two filter papers into the bottle and replace the cap. Weigh the bottle and papers to the nearest 0.1 mg. and record the weight.

Iontophoresis and Sweat Collection

1. Remove two filter paper discs from the stock box of filter papers with forceps and moisten them thoroughly with the pilocarpine solution. Superimpose the two papers and apply them to the site to be sweated. This is usually the flexor surface of the forearm, but in the case of very young children the forehead or the right scapula may be preferred.

2. Apply the positive electrode (red lead) to the pilocarpine-saturated paper discs so that the metal is in contact with the paper only and does not touch the skin. Secure the electrode with a rubber strap.

3. Remove two more filter papers from the box and soak these in isotonic saline solution. Apply these to the extensor surface of the forearm (forearm sweating) or to the right upper arm (forehead and scapula sweating).

[4] Marathon Division of American Can Company, Menasha, Wisconsin.
[5] "Kimwipes," Kimberly-Clark Corporation, Neenah, Wisconsin, are satisfactory.

4. Apply the negative electrode (black lead) to the saline-soaked papers and secure with a rubber strap.

5. Switch on the power supply and slowly increase the current to 0.5 ma. Wait for a few seconds and then increase the current to 2 ma. and maintain for 5 minutes. Adjust the current to 2 ma. if it has an initial tendency to drift. Do not leave the patient during this time. Be certain to investigate if the patient shows any sign of discomfort. Should this occur, the most usual causes are: (a) one of the electrodes has slipped and is in contact with the skin, (b) the filter papers are not adequately saturated with their respective solutions, or (c) the patient is unusually sensitive to the procedure. In either of the first two cases, switch off the power supply and make the necessary adjustment. When the patient seems to be especially sensitive, try lowering the current to 1.5 ma.

NOTE: The iontophoresis apparatus *must* be used with great care if burns are to be avoided (J. M. W.).

6. After 5 minutes of iontophoresis, reduce the current to 0 and turn off the power supply.

7. Wash the sweating site three times with tissues soaked with deionized water and then dry the area with paper tissues.

8. Hold the weighing bottle with tissues and remove the cap. Remove the filter papers from the bottle with forceps and place them in the center of the area of sweating. Discard the protective paper layer from a square of Parafilm and cover the filter papers with the clean surface. Seal the four edges of the Parafilm to the skin with masking tape so that the seal is airtight.

9. Allow the filter papers to remain in place for at least 30 minutes and preferably for 1 hour.

10. Using tissues, remove the cap from the weighing bottle. Release two edges of the adhesive tape, remove the filter papers with forceps, and return them to the bottle. Cap it immediately.

11. Reweigh the bottle containing the sample as soon as possible and calculate the weight of sweat by the increase in weight.

NOTE: Although 50 mg. of sweat is often accepted as the minimum quantity for an accurate chloride determination, it is preferable to raise this to 100 mg. in order to decrease the possibility of error (J. M. W.).

I. CHLORIDE DETERMINATION BY COTLOVE
AMPEROMETRIC TITRATION METHOD (14)[6]

Reagents

1. *Acid diluent.* To 900 ml. deionized water add 6.4 ml. concentrated nitric acid and 100 ml. glacial acetic acid. Mix and store in a glass container at room temperature. This reagent is stable indefinitely.

2. *Gelatin reagent.* Weigh 6.0 g. gelatin (Knox Unflavored No. 1),[7] 0.1 g. thymol blue (water soluble),[8] and 0.1 g. thymol. Add approximately 1 l. of hot deionized water and heat gently until the solution is clear. Cool to room temperature and dispense into test tubes, each sufficient for a day's use. Store under refrigeration. The reagent is stable for at least 6 months. Use a fresh tube each day and discard the unused portion.

3. *Stock chloride standard, 100 meq./l.* Dissolve 5.845 g. of purest grade anhydrous sodium chloride in deionized water and dilute to 1 l. Store at room temperature in a polyethylene container.

4. *Working chloride standard, 0.5 meq./l.* Dilute the stock chloride standard 1:200 with the acid diluent, reagent 1. Store at room temperature. Four ml. of this standard solution contains 0.002 meq. of chloride.

Procedure

1. Add 5.00 ml. acid diluent (reagent 1) to the plastic bottle containing the sweat sample and shake thoroughly.

2. Set up 3 titration vials as follows:

> Blank: 4.00 ml. acid diluent (reagent 1)
> Standard: 4.00 ml. dilute standard (reagent 4)
> Test: 4.00 ml. aliquot from the bottle containing the sweat sample

3. Add to each vial 0.15 ml. of gelatin reagent.

4. Titrate on the Cotlove titrating apparatus with the selector switch in the medium position.

---

[6] Refer to Volume 3 of "Standard Methods of Clinical Chemistry" (15).

[7] Chas. B. Knox Gelatin Company, Inc., Johnstown, New York.

[8] National Aniline Division, Allied Chemical Corporation, 40 Rector Street, New York 6, New York.

Calculation

$$\text{meq./l.} = 0.002 \times \frac{T - B}{S - B} \times \frac{5 + W}{4} \times \frac{1000}{W} = \frac{T - B}{S - B} \times \frac{5 + W}{2W}$$

T = titration of test
S = titration of standard
B = titration of blank
W = weight in grams of sweat collected

NOTE: Run a reagent blank each time a new box of filter papers is used to ensure that the batch is chloride-free. This is done by processing two filter paper discs through steps 1 to 4 of the above procedure (M. J. R.).

II. CHLORIDE DETERMINATION BY SCHALES AND SCHALES TITRIMETRIC METHOD (MICRO) (16)

Reagents

1. Mercuric nitrate, stock solution, approximately 0.01 N. Pipet 20.0 ml. of 2 N nitric acid into a 1 l. flask containing about 500 ml. deionized water. Add 1.85–1.90 g. of mercuric nitrate, $Hg(NO_3)_2 \cdot H_2O$, and mix until dissolved. Make up to 1 l. with deionized water and mix. The solution is stable indefinitely.

2. Mercuric nitrate, working solution, approximately 0.001 N. Dilute exactly 1 vol. of reagent 1 with 9 vol. of 0.04 N nitric acid (reagent 6).

3. Indicator solution Dissolve 50 mg. of s-diphenylcarbazone in 50 ml. of 95% ethanol. Store in a brown bottle in a refrigerator. The reagent is stable for about 1 month if protected from light at all times.

4. Chloride standard, 10 meq./l. Dilute the 100 meq./l. stock chloride standard 1:10 with deionized water.

5. Nitric acid, approximately 0.01 N. Dilute 0.60 ml. concentrated nitric acid to 1 l. with deionized water.

6. Nitric acid, approximately 0.04 N. Dilute 2.50 ml. concentrated nitric acid to 1 l. with deionized water.

Procedure

1. Add to the plastic bottle containing the sweat sample 1.00 ml. of deionized water and shake thoroughly.

2. Pipet 0.20 ml. of standard, water, and diluted sweat into separate 15-ml. beakers.

3. Add 1.0 ml. 0.01 N nitric acid to each beaker.

4. Add approximately 0.1 ml. of indicator solution to each beaker.

5. Titrate the contents of each beaker with 0.001 N mercuric nitrate (reagent 2), using a 5 ml. buret graduated in 0.01 ml., to the first appearance of a pale violet color.

Calculation

$$\text{meq./l.} = 0.002 \times \frac{T - B}{S - B} \times \frac{1 + W}{0.2W} \times 1000 = 10 \times \frac{T - B}{S - B} \times \frac{1 + W}{W}$$

T = titration of test
S = titration of standard
B = titration of blank
W = weight in grams of sweat collected

NOTE: The above version of the titrimetric procedure was suggested for use in those laboratories possessing neither the Cotlove chloride titrating apparatus nor ultramicro equipment (J. M. W.).

III. CHLORIDE DETERMINATION BY SCHALES AND SCHALES TITRIMETRIC METHOD (ULTRAMICRO) (16)

Reagents

1. *Mercuric nitrate, approximately 0.1 N.* Dissolve 1.6 g. mercuric nitrate in about 70 ml. of deionized water containing 1 ml. of concentrated nitric acid. Make up to 100 ml. with water.

2. *Indicator solution.* Prepare as described in the previous section.

3. *Nitric acid, approximately 0.03 N.* Dilute 0.20 ml. of concentrated nitric acid to 100 ml. with deionized water.

4. *Chloride standard, 3.00 meq./l.* Dilute 100 meq./l. stock chloride standard 3:100 with 0.03 N nitric acid.

Apparatus

1. *Ultramicro buret capable of measuring to 0.01 μl. or less.*[9,10]

2. *Micro titration cups.*[9]

3. *Pipets, 80 μl. calibrated to deliver.* These may be of the Lang-Levy pattern or the Beckman/Spinco polyethylene design.[9]

[9] Beckman/Spinco No. 153 Microtitrator, Beckman/Spinco Division, 1117 California Avenue, Palo Alto, California.

[10] Gilmont Ultramicro Buret, A. H. Thomas Company, Philadelphia 5, Pennsylvania.

Procedure

1. Add 1.00 ml. of 0.3 N nitric acid to the plastic bottle containing the sweat sample and shake thoroughly.
2. Pipet 80 μl. of dilute sweat, standard, and 0.03 N nitric acid, into separate titration cups.
3. Add 80 μl. of indicator solution to each cup.
4. Titrate each solution to the pale violet end point with 0.1 N mercuric nitrate reagent.

Calculation

$$\text{meq./l.} = \frac{0.16}{1000} \times \frac{T - B}{S - B} \times \frac{1 + W}{0.08W} \times 1000 = 2 \times \frac{T - B}{S - B} \times \frac{1 + W}{W}$$

T = titration of test
S = titration of standard
B = titration of blank
W = weight in grams of sweat collected

Normal and Abnormal Values

The following values relate to sweat collected after iontophoresis with pilocarpine.

NORMAL (11, 17, 18)

Ninty-six per cent of normal children have sweat chloride values of less than 30 meq./l., and the remainder are 60 meq./l. or less.

ABNORMAL (11, 17, 18, 19, 20)

Values for patients with cystic fibrosis ranging from 40 to 154 meq./l. have been reported, whereas sweat chloride values of relatives of patients with this disease were within the normal range. In eight children with glycogen storage disease Type 1, the range extended up to 109 meq./l. Levels of 50–60 meq./l. have been found in one case of pitressin-resistant diabetes insipidus. In two series of children with allergy, sweat chloride values were similar to those obtained from normals.

NOTE: Values obtained using techniques other than pilocarpine iontophoresis are as follows (9, 10, 21, 22, 23):
Sweat obtained without the application of an external stimulus from children

in a control group and in a group with chronic pulmonary disease showed chloride concentrations of 38.4 meq./l. or less. In patients with cystic fibrosis, however, levels were found up to 196.1 meq./l.

Use of plastic bag method of sweat stimulation gave a mean value for normal subjects of 20.9 meq./l. and 116.9 meq./l. for patients with cystic fibrosis. Sweat obtained by this technique from two brothers aged 18 and 22 years, both suffering from familial pulmonary disease, gave chloride concentrations in the range of 60–117 meq./l.

In an investigation of 262 adults, the reported normal values ranged from 4.1 to 68.2 meq./l. Patients from an outpatient chest clinic (medical) showed the highest levels of sweat chloride, 16.3–125.9 meq./l., excluding patients with cystic fibrosis, who had levels exceeding 75 meq./l.

After Mecholyl induction of sweating, results indicated that there was a small increase in sweat chloride levels in children as age increased. Adults from normal families and families which included patients with cystic fibrosis showed values greater than 60 meq./l. in 30% of the assays. For this reason, the test is regarded as of limited usefulness in adolescents and adults. Results from men subjected to a uniform thermal stimulus indicated that a group with chronic pulmonary disease had a sweat chloride concentration significantly greater than a comparable control group.

REFERENCES

1. di Sant'Agnese, P. A., Darling, R. C., Perera, G. A., and Shea, E., Sweat electrolyte disturbances associated with childhood pancreatic disease. Am. J. Med. 15, 777–784 (1953).
2. Darling, R. C., di Sant'Agnese, P. A., Perera, G. A., and Andersen, D. H., Electrolyte abnormalities of the sweat in fibrocystic disease of the pancreas. Am. J. Med. Sci. 225, 67–70 (1953).
3. di Sant'Agnese, P. A., Darling, R. C., Perera, G. A., and Shea, E., Abnormal electrolyte composition of sweat in cystic fibrosis of the pancreas. Pediatrics 12, 549–563 (1953).
4. Schwachman, H., Leubner, H., and Catzel, P., Mucoviscidosis. Advan. Pediat. 7, 249–323 (1955).
5. Misch, K. A., and Holden, H. M., Sweat test for the diagnosis of fibrocystic disease of the pancreas. Report of a fatality. Arch. Disease Childhood 33, 179–180 (1958).
6. Mielke, J. E., and Durrance, F. Y., Sweat electrolytes in familial pulmonary disease. Diseases of Chest 39, 601–608 (1961).
7. Dubowski, K. M., Some practical simplifications of perspiration electrolyte analysis ("sweat test"). Clin. Chem. 7, 494–503 (1961).
8. Gorvoy, J. D., Acs, H., and Stein, M. L., The hazard of induction of sweating in cystic fibrosis of the pancreas. Pediatrics 25, 977–982 (1960).
9. Weeks, M. M., and Brown, G. A., Sweat analysis in fibrocystic disease, chronic pulmonary disease, and controls. Arch. Diseases Childhood 33, 74–77 (1958).
10. Anderson, C. M., and Freeman, M., A simple method of sweat collection with analysis of electrolytes in patients with fibrocystic disease of the pancreas and their families. Med. J. Australia 1, 419–422 (1958).

11. Gibson, L. E., and Cooke, R. E. A test for concentration of electrolytes in sweat in cystic fibrosis of the pancreas utilizing pilocarpine iontophoresis. *Pediatrics* 23, 545–549 (1959).

12. Schwachman, H., and Gahm, N., Studies in cystic fibrosis of the pancreas. A simple test for the detection of excessive chloride on the skin. *New Engl. J. Med.* 255, 999–1001 (1956).

13. MacFarlane, J. C. W., Norman, A. P., and Stroud, C. E., Fingerprint sweat test in fibrocystic disease of pancreas. *Brit. Med. J.* II, 274–275 (1957).

14. Cotlove, E., Trantham, H. V., and Bowman, R. L., An instrument and method for automatic rapid, accurate, and sensitive titration of chloride in biologic samples, *J. Lab. Clin. Med.* 51, 461–468 (1958).

15. Cotlove, E., Chloride. *In* "Standard Methods of Clinical Chemistry" (D. Seligson, ed.), Vol. 3, pp. 81–92. Academic Press, New York, 1961.

16. Schales, O., Chloride. *In* "Standard Methods of Clinical Chemistry" (M. Reiner, ed.), Vol. 1, pp. 37–42. Academic Press, New York, 1953.

17. Harris, R. C., and Cohen, H. I., Sweat electrolytes in glycogen storage disease Type 1. *Pediatrics* 31, 1044–1046 (1963).

18. Andrews, B. F., Bruton, O. C., and Knoblock, E. C., Sweat chloride concentration in children with allergy and with cystic fibrosis of the pancreas. *Pediatrics* 29, 204–208 (1962).

19. Lobeck, C. C., Barta, R. A., and Mangos, J. A., Study of sweat in pitressin-resistant diabetes insipidus. *J. Pediat.* 62, 868–875 (1963).

20. Perry, E. F., and Scott, R. B., Sweat electrolytes in allergic children. *J. Allergy* 32, 528–530 (1961).

21. Schwachman, H., Dooley, R. R., Guilmette, F., Patterson, P. R., Weil, C., and Leubner, H., Cystic fibrosis of the pancreas with varying degrees of pancreatic insufficiency. *A.M.A. J. Diseases Children* 92, 347–368 (1956).

22. Toigo, A., Dietz, A. A., Crane, J. B., and Reisner, D., Sweat electrolytes in chronic pulmonary disease. *Ann. Internal Med.* 58, 961–968 (1963).

23. Peterson, E. M., Consideration of cystic fibrosis in adults, with a study of sweat electrolyte values. *J. Am. Med. Assoc.* 171, 1–6 (1959).

ULTRAMICRO GLUCOSE (ENZYMATIC)*

Submitted by: SAMUEL MEITES, The Children's Hospital, Columbus, Ohio
Checked by: E. GREGORY ASHE, The Presbyterian Hospital, Denver, Colorado
CARL E. MOYER, Parke-Davis Research Laboratories, Ann Arbor, Michigan
EUGENE W. RICE, William H. Singer Memorial Research Laboratory, Allegheny General Hospital, Pittsburgh, Pennsylvania

Introduction

Although the enzymatic determination of glucose was comprehensively treated in Volume 4 of "Standard Methods of Clinical Chemistry" (1), two general questions remain to be considered. How should the blood of small children be preserved prior to analysis? What micro method is best suited for blood which, in most instances, must be collected by "capillary" puncture? These particular questions arise from the fact that the newborn, whose blood may be markedly hypoglycemic (by adult standards), demonstrates a considerably increased rate of glycolysis over the blood of the adult (2, 3). The usual preservatives are inconvenient to use when capillary blood is routinely obtained. Moreover, it is desirable to employ ultramicro volumes of blood for analysis because the number of multiple procedures requested per pediatric patient continues to rise.[1]

A simple technique for blood preservation has been coupled with an ultramicro procedure for glucose analysis (3). Aqueous dilution of whole blood, which is readily performed at the bedside, will stabilize glucose for at least 2 hours. However, the ultramicro analysis of blood diluted to the degree needed for its glucose preservation (1 + 80) requires a method of high sensitivity. The Somogyi and Nelson technique (4, 5, 6, 7) does not meet this requirement, whereas

---

* Based on the modification of Washko and Rice (8).

[1] In 1961, about 23,000 blood chemistry tests, exclusive of endocrine assays, coagulation studies, and procedures in other sections of the clinical laboratory, were made on 9,500 patients (principally infants) at The Children's Hospital, Columbus, Ohio. Fifty-three per cent of the patients received multiple tests (the average per patient was four tests per 24 hours) representing 82% of all the blood chemistry performed. Twenty per cent of all the tests done were on patients for whom at least *six* chemical procedures were carried out per day.

113

the glucose oxidase procedure is suitable with minor modification (8, 9).

Principles

The glucose concentration in deproteinized blood is determined indirectly with a coupled enzyme system, at pH 7.0, as follows:

(1) Glucose $+ O_2 + H_2O \xrightarrow{\text{glucose oxidase}} H_2O_2 +$ gluconic acid

(2) $H_2O_2 + o$-dianisidine $\xrightarrow{\text{peroxidase}} H_2O +$ oxidized o-dianisidine

The oxidized o-dianisidine is converted to a red pigment by the addition of strong mineral acid, and the color produced is proportional in intensity to the glucose concentration. The color is measured spectrophotometrically. A more detailed explanation of the principles in this method may be obtained from Volume 4 of this series (1).

Reagents

1. *Phosphate buffer, 0.08 M, pH 7.0.* Dissolve 6.96 g. of anhydrous Na_2HPO_4 and 4.24 g. of anhydrous KH_2PO_4 in water and make up to a final volume of 1 l.

2. *Buffered glycerol.* Mix 600 ml. of phosphate buffer with 400 ml. of glycerol. This reagent is stable for several months if stored at 4°C.

3. *Glucose oxidase reagent (GOR).* Place 250 mg. of glucose oxidase[2] and 10 mg. of peroxidase[3] in a 100 ml. graduated cylinder containing about 60 ml. buffered glycerol. Add 2.0 ml. of 1% (w/v) o-dianisidine[4] in 95% ethanol (or absolute methanol) and dilute to 100 ml. with buffered glycerol. Mix thoroughly and filter through a good grade of fluted filter paper, such as Whatman No. 42. The reagent is stable for at least 2 weeks if stored in a brown bottle at less than 4°C. The chromogen is sensitive to light and darkens on standing.

NOTE: The GOR has twice the concentration of that used in the macro procedure (8).

[2] Glucose oxidase, fungal, approx. 1500 units per gram, Sigma Chemical Co., 3500 DeKalb St., St. Louis 18, Missouri.
[3] Horseradish peroxidase, Grade D, Worthington Biochemical Corp., Freehold, New Jersey.
[4] Eastman No. P509, 3,3'-dimethoxybenzidine, Distillation Products Industries, Rochester 3, New York.

NOTE: The enzyme and chromogen components of this reagent are supplied in 2 separate vials under the trademark "Glucostat."[5] To prepare 100 ml. of GOR, add 1 ml. of buffered glycerol to each of 2 chromogen (o-dianisidine) vials. Transfer to a 100 ml. graduated cylinder with several rinsings of buffered glycerol. Dissolve the contents of each of 2 Glucostat (glucose oxidase and peroxidase) vials in buffered glycerol and transfer, with rinsing, to the cylinder. Dilute to the mark with buffered glycerol, then filter and store as described previously.

4. *Sulfuric acid, approx. 7.6 N.* Mix 1 vol. of concentrated H_2SO_4 (sp. gr. 1.84) with 4 vol. of H_2O.

5. *Sulfuric acid, approx. 5 N.* Mix 2 vol. of 7.6 N H_2SO_4 with 1 vol. of H_2O.

6. *Glucose stock standard, 1000 mg./100 ml. (w/v) in 0.25% (w/v) benzoic acid.* This reagent is stable at room temperature.

7. *Glucose working standards.* Prepare by dilution of the stock standard with 0.25% (w/v) benzoic acid. The working standards are stable at room temperature for at least 2 months.

8. *Deproteinizing reagents, Somogyi (5).*

(a) *Barium hydroxide, 0.3 N.* Dissolve 47.3 g. of $Ba(OH)_2$ in water and dilute to 1 l. Store the solution in a paraffin-lined or polyethylene bottle containing a soda lime tube and siphon.

(b) *Zinc sulfate, 5%.* Dissolve 50 g. of $ZnSO_4 \cdot 7H_2O$ in water and dilute to 1 l.

(c) Adjust the two solutions with water so that 10.00 ml. of the zinc sulfate solution, diluted with about 30 to 40 ml. of water, is titrated to a faint pink color by 10.00 ± 0.05 ml. of barium hydroxide, in the presence of phenolphthalein as indicator.

NOTE: Sodium hydroxide, 0.3 N (made by dissolving 12 g. of NaOH in water and diluting to 1 l., then titrating as indicated in step c), may be substituted for $Ba(OH)_2$ (10). Filtrates made from $Ba(OH)_2$, however, may also be used for plasma and serum glucose determination. In addition, salts from these protein-precipitating reagents are not introduced into the supernatant fluid, and anticoagulants such as fluoride and oxalate are removed (5).

Procedure

1. Pipet the following volumes, in the order given, into 12 × 75 mm. test tubes:

[5] Glucostat—25 micro tests per set. Worthington Biochemical Corp., Freehold, New Jersey.

	Unknown (ml.)	Standard (ml.)	Blank (ml.)
Water	2.00	2.00	2.00
Blood	0.025	—	—
Standard (200 mg./100 ml.)	—	0.025	—
Ba(OH)$_2$	0.10	0.10	0.10
ZnSO$_4$	0.10	0.10	0.10

NOTE: The water may be predispensed into the test tubes and the tubes sealed with Parafilm. They may then be transported in a collecting tray.

NOTE: Puncture a fingertip, heel, or big toe and wipe away the first 2 drops. Use the next few drops to fill the pipet. A gentle milking action is permitted at a distance as remote as possible from the puncture site. Excessive squeezing, though difficult to define, will cause erratic glucose values. For collecting and measuring the blood, use a TC micropipet, preferably with A.C.S. specifications (11). Once the aqueous dilution is made, the blood is stable for 2 hours.

Allow the tubes to stand for 5 minutes after the addition of hydroxide. Mix after each addition.

NOTE: Follow the order of addition of reagents strictly. When NaOH replaces Ba(OH)$_2$, add the ZnSO$_4$, first (10). Reversal of the order of addition or pre-mixing of reagents may lead to slight, but significant error.

2. Cetrifuge for 5 minutes at 2000–3000 r.p.m.

3. Pipet 1.00 ml. of each supernatant fluid into appropriately labeled cuvets or test tubes.

4. Add 1.00 ml. of GOR to each container and mix by gently inverting or tapping.

5. Incubate for 30 minutes at 37°C. in a water bath.

6. Add 4.00 ml. of 7.6 N H$_2$SO$_4$ and mix by tapping.[6] A pink color develops immediately and remains stable for many hours.

7. Set the blank at zero absorbance and 540 mμ, and read the absorbance of the unknown and standard.

Calculation

$$\text{mg. glucose/100 ml.} = \frac{A_{unk}}{A_{std}} \times \text{mg. glucose per 100 ml. standard}$$

NOTE: The concentration-absorbance curve is linear to 400 mg./100 ml. provided the reaction mixture of step 6 is diluted with an equal volume (6 ml.) of 5 N H$_2$SO$_4$, when the level of glucose is above 250 mg./100 ml.[7]

[6] A Hycel Reagent Dispenser, with a 4 ml. chamber calibration, and a 400 ml. reservoir, is useful for this step.

[7] These results were found with the Coleman Junior Model 6-A spectro-

Discussion

When blood is taken from infants by venipuncture the use of an ultramicro volume for analysis is unnecessary. If, on the other hand, blood cannot always be obtained by venipuncture one should have available ultramicro procedures adapted to the volumes obtainable from capillary puncture. The practical experience of several micro-chemists indicates that the use of 50 μl. or less for glucose analysis leaves sufficient blood for other microchemical procedures.

It is unwise to assume, without experimental evidence, that an ultramicro procedure may be safely established by the "scaling-down" process as suggested in Volume 4 of "Standard Methods of Clinical Chemistry" (1). The method presented here is not only ultramicro, but employs a unique way of protecting glucose from the glycolytic mechanism for at least 2 hours. Consequently, it eliminates: special tubes containing preservatives, rapid centrifugation to remove serum or plasma for analysis, the use of an ice bath, or the immediate addition of protein precipitants.

Glycolysis in leucocytes may cause a factitious hypoglycemia in shed blood when the leucocyte count is elevated. This is particularly true for leukemic patients whose blood may not be preserved completely by the addition of sodium fluoride (12).

NOTE: In the Submitter's laboratory, the aqueous dilution technique has effectively prevented loss of glucose for 2 hours in the blood of three leukemic patients with white blood counts of 38,000, 60,000 and 576,000. Losses of glucose from the same specimens, undiluted, were 29%, 26%, and 56%, respectively.

The standard deviation of the ultramicro method was ±2.8 for 45 consecutive determinations of a 200 mg./100 ml. glucose standard (9). This compared favorably with the value of ±1.8 obtained for a 100 mg./100 ml. standard in the original macro procedure (8). When 25 μl. of a blood specimen containing 79 mg./100 ml. glucose and 25 μl. of each of a series of standard glucose solutions (50 to 300 mg./100 ml.) were added to 2.00 ml. of water, the recoveries ranged from 97 to 102% (9). Because of these excellent recoveries, the order of addition of Ba(OH)$_2$ and ZnSO$_4$ remains according to

photometer. Checker (E. G. A.) reports the upper limit of glucose concentration at which dilution is essential is 350 mg./100 ml. for the Klett-Summerson photoelectric colorimeter (filter No. 54).

Somogyi's original method (5), and has not been attempted in reverse as recommended for NaOH and $ZnSO_4$ (10).

NOTE: Checker (C. E. M.) reports the following precision studies: *The mean, range of values,* and *standard deviation* obtained by the ultramicro method from 20 replicate determinations of a "control" serum with a stated value of 103 mg./100 ml. were 104.5 mg./100 ml., 99–110 mg./100 ml., and ± 2.5. The corresponding values with the Somogyi-Nelson method were 103.3 mg./100 ml., 97–110 mg./100 ml., and ± 3.9. The results obtained from the ultramicro analysis of a control serum of 195 mg./100 ml. were 194.8 mg./100 ml., 188–203 mg./100 ml., and ± 4.9. The Somogyi-Nelson technique gave 194.1 mg./100 ml., 183–205 mg./100 ml., and ± 7.1. The mean differences obtained from duplicate analysis of 15 blood specimens with glucose concentrations ranging from 70 to 212 mg./100 ml. were ± 2.5 mg. for the ultramicro method and 7.6 mg. for the Somogyi-Nelson method.

There are numerous *potential* sources of interference with the glucose oxidase method, but few of these have been evaluated (1). The potential interfering effects of bilirubin and hemolysis are at least partially removed during protein precipitation although convincing data have not been reported (13).

This method is highly specific for glucose among the physiologically important sugars. This specificity may be used to advantage in tolerance tests for reducing sugars other than glucose based on differences obtained between the Somogyi-Nelson technique (which measures nonglucose reducing sugars as well as glucose), and the enzymatic method (which measures glucose alone) (14, 15, 16, 17).

Although the method may lend itself well to automatic analysis, this has not, as yet, been reported. Dilutions of blood made at the bedside could be used satisfactorily.

The method is applicable to cerebrospinal fluid. However, 0.05 ml. rather than 0.025 ml. of specimen should be used because of the reduced concentration of glucose. Final glucose values in cerebrospinal fluid would then be divided by two.

Normal Levels

A meaningful discussion of the normal blood sugar levels in man is limited by several variables not completely controlled in the past. These include: (1) physiological variations in subjects due to age, sex, diurnal rhythms, stress, season, and previous diet and exercise; (2) variation in the specificity of the analytical methods used; (3) variation in the specimen analyzed—whether whole blood, serum

TABLE I

NORMAL VALUES OF GLUCOSE IN CAPILLARY WHOLE BLOOD[a]

Subjects	Age	No. of subjects tested	Method of analysis	Mean (mg./ 100 ml.)	Range (mg./ 100 ml.)
Prematures (18)	1 day	37	Glucose oxidase	45	23–84
	1 week	33	Glucose oxidase	43	22–83
	1 month	52	Glucose oxidase	52	18–77
	1–2 months	43	Glucose oxidase	48	22–83
Full term newborns (19)	1 day	51	Somogyi-Nelson	57	40–73
	1 week	51	Somogyi-Nelson	67	55–80
Adults (20)		36	Glucose oxidase	69	46–94

[a] Capillary blood sugar values may be 20 to 50 mg./100 ml. greater than venous values after a test dose of glucose is administered to a subject (21).

or plasma, and whether obtained from venous, arterial, or "capillary" source; and (4) variation in preservation techniques.

Table I is presented as a guide to a few normal values of glucose in the properly preserved capillary whole blood of infants in the postabsorptive state. Emphasis is placed on *age* as a variable.

REFERENCES

1. Fales, F. W., Glucose (enzymatic). In "Standard Methods of Clinical Chemistry" (D. Seligson, ed.), Vol. 4, pp. 101–112. Academic Press, New York, 1963.
2. Baens, G. S., Oh, W., Lundeen, E., and Cornblath, M., Determination of blood sugar in newborn infants. *Pediatrics* 28, 850–851 (1961).
3. Meites, S., and Bohman, N., In vitro stabilization of blood glucose with water. *Clin. Chem.* 9, 594–599 (1963).
4. Somogyi, M., A new reagent for the determination of sugars. *J. Biol. Chem.* 160, 61–68 (1945).
5. Somogyi, M., Determination of blood sugar. *J. Biol. Chem.* 160, 69–73 (1945).
6. Nelson, N., A photometric adaptation of the Somogyi method for the determination of blood sugar. *J. Biol. Chem.* 153, 375–380 (1944).

7. Somogyi, M., Notes on sugar determination. *J. Biol. Chem.* **195**, 19–23 (1952).
8. Washko, M. E., and Rice, E. W., Determination of glucose by an improved enzymatic procedure. *Clin. Chem.* **7**, 542–545 (1961).
9. Meites, S., and Bohman, N., Evaluation of an ultramicro method for glucose determination. *Am. J. Med. Technol.* **29**, 327–331 (1963).
10. Welch, N. L., and Danielson, W. H., The effect of different methods of precipitation of protein on the enzymatic determination of blood glucose. *Am. J. Clin. Pathol.* **38**, 251–255 (1962).
11. Steyermark, A., Alber, H. K., Aluise, V. A., Huffman, E. W. D., Jolley, E. L., Kuck, J. A., Moran, J. J., and Ogg, C. L., Report on recommended specifications for microchemical apparatus. Volumetric glassware. Microliter pipets. *Anal. Chem.* **30**, 1702–1703 (1958).
12. Hanrahan, J. B., Sax, S. M., and Cillo, A., Factitious hypoglycemia in patients with leukemia. *Am. J. Clin. Pathol.* **40**, 43–45 (1963).
13. Saifer, A., and Gerstenfeld, S., The photometric microdetermination of blood glucose with glucose oxidase. *J. Lab. Clin. Med.* **51**, 448–460 (1958).
14. Tygstrup, N., Winkler, K., Lund, E., and Engell, H. C., A clinical method for determination of plasma galactose in tolerance tests. *Scand. J. Clin. Lab. Invest.* **6**, 43–48 (1954).
15. Sondergaard, G., Micro-method for determination of blood galactose by means of glucose oxidase (notatin) and anthrone. *Scand. J. Clin. Lab. Invest.* **10**, 203–210 (1958).
16. Behrendt, H., "Diagnostic Tests in Infants and Children," 2nd ed., p. 50. Lea & Febiger, Philadelphia, Pennsylvania, 1962.
17. Meites, S., and Faulkner, W. R., "Manual of Practical Micro and General Procedures in Clinical Chemistry," p. 186. Thomas, Springfield, Illinois, 1962.
18. Baens, G. S., Lundeen, E., and Cornblath, M., Studies of carbohydrate metabolism in the newborn infant. VI. Levels of glucose in blood in premature infants. *Pediatrics* **31**, 580–589 (1963).
19. Norval, M. A., Kennedy, R. I. J., and Berkson, J., Blood sugar in newborn infants. *J. Pediat.* **34**, 342–351 (1949).
20. Middleton, J. E., Experience with a glucose oxidase method for estimating glucose in blood and c.s.f. *Brit. Med. J.* **I**, 824–826 (1959).
21. Joslin, E. P., Root, H. F., White, P., and Marble, A., "The Treatment of Diabetes Mellitus," pp. 158–169. Lea & Febiger, Philadelphia, Pennsylvania, 1959.

REVIEW

22. Free, A. H., Enzymatic determinations of glucose. *Advan. Clin. Chem.* **6**, 67–96 (1963).

LEAD IN BLOOD AND URINE*

Submitted by: Eugene W. Rice, William H. Singer Memorial Research Labora-
tory, Allegheny General Hospital, Pittsburgh, Pennsylvania
Dean C. Fletcher, University of Nevada Desert Research In-
stitute, and Washoe Medical Center, Reno, Nevada, and
Agnes Stumpff, St. Mary's Hospital, Reno, Nevada
Checked by: Robert B. Foy, Edward W. Sparrow Hospital, Lansing, Michigan
Irving Sunshine, Institute of Pathology, Western Reserve Uni-
versity, Cleveland, Ohio

Introduction

Throughout the years much effort has been directed toward the development of satisfactory photometric procedures for the determination of the trace amounts of lead present in urine and blood. Most investigators have utilized the sensitive color reaction for lead based on the formation of a red chloroform-soluble complex with diphenylthiocarbazone (dithizone). With the advent of more recent refinements (1, 2, 3), analyses are now more readily performed on a routine basis, thereby expediting the clinical diagnosis and study of lead poisoning.

The present dithizone procedure represents a number of modifications designed to shorten the time of analysis, lessen the danger of contamination, and reduce the quantity of blood or urine customarily required. Use of a single tube for the digestion and extraction obviates the need for quantitative transfers of the solutions before the excess dithizone has been removed.

Principle

Lead forms a red chloroform-soluble complex with diphenylthiocarbazone (dithizone). Interference from other metals, such as copper and iron, is eliminated by extraction of the lead-dithizone complex from alkaline solution (pH > 11) in the presence of cyanide. The unreacted dithizone is removed with ammonium hydroxide-cyanide solution and the red complex remaining in the chloroform layer is determined spectrophotometrically at a wavelength of 510 mμ. The procedure, with the exception of the spectrophotometry, is performed

* Based on the method of Bessman and Layne (1).

121

in a single glass-stoppered digestion tube, thereby eliminating the use of separatory funnels and the usual transfer steps. The basic color reaction is represented as follows:

$$Pb^{2+} + 2S=C \underset{NH-NH}{\overset{N=N}{\diagup}} \xrightarrow{pH > 11} S=C \underset{NH-N-Pb-N-NH}{\overset{N=N}{\diagup}} \overset{N=N}{\diagup} C=S + 2H^+$$

Dithizone
(green in chloroform)

Lead-dithizone complex
(red in chloroform)

Reagents

NOTE: It is imperative that all glassware be washed with detergent and water, soaked overnight in dilute nitric acid (1 + 1) followed by rinsing in hot tap water, and finally with generous portions of distilled-deionized water. Glassware thus cleaned should be reserved exclusively for trace metal determinations and must be protected carefully from contamination. Either distilled-resin deionized water or "glass double-distilled water" should be employed for all reagents. Reagent grade chemicals are used unless otherwise noted.[1]

1. *Digestion reagent.* Cautiously mix 25 vol. of concentrated nitric acid[1] and 10 vol. of concentrated sulfuric acid.[1] Prepare fresh reagent for each set of determinations.

2. *Perchloric acid, 70% (w/v).*[1] Store in the refrigerator.

3. *Ammonium hydroxide, concentrated.*[1]

[1] Bessman and Layne and Checker (I. S.) use reagent grade compounds, and do not recommend further purification. Lower reagent blanks may be obtained, however, by employing the following purified reagents available from G. Frederick Smith Chemical Company, 867 McKinley Avenue, Station D, Box 5906, Columbus 22, Ohio:

Double vacuum distilled sulfuric acid, 95% H_2SO_4 (Item No. 273).
Redistilled concentrated HNO_3 (Item No. 63).
Double vacuum distilled perchloric acid, 70% w/v $HClO_4$ (Item No. 67).
Ammonium hydroxide, iron-free, specific gravity 0.92 (Item No. 108–11).
Dithizone (Item No. 139).
Chloroform-redistilled, iron and copper free (Item No. 108–16).

4. Phenol red, 0.10% (w/v). Store the solution in a plastic dropping bottle.

5. Dithizone solution. Dissolve 30.0 mg. of diphenylthiocarbazone (dithizone)[1] in exactly 1 l. of redistilled chloroform.[1] Store this solution in a brown bottle in a refrigerator, but allow it to warm to room temperature before using.

NOTE: In order to prepare redistilled chloroform, dissolve 0.5 g. of dithizone in 2 l. of chloroform, distill in an all-glass apparatus, discarding the first 100 ml. of distillate and collecting approximately 1800 ml.

6. Buffer solution. Dissolve 118 g. of dibasic ammonium citrate, $(NH_4)_2HC_6H_5O_7$, in water, add 75 ml. of concentrated ammonium hydroxide, and dilute to 250 ml. with water. When the solution has cooled to room temperature, add 5.0 g. of potassium cyanide, KCN (*extremely poisonous!*), and 2.5 g. of anhydrous sodium sulfite, Na_2SO_3. Extract the buffer solution with dithizone solution until the chloroform layer remains green. Add 500 ml. of concentrated ammonium hydroxide. This solution is stable at room temperature.

NOTE: Perform this entire procedure in a glass-stoppered borosilicate bottle marked on the outside with lines corresponding to the volumes to which the solution is diluted. Accomplish the extraction by adding approximately 2-ml. aliquots of dithizone and shaking the stoppered bottle. Aspirate the chloroform-dithizone layer from beneath the aqueous layer with a pipet and rubber bulb, and place it in a test tube through which air is bubbled for a few seconds in order to remove ammonia. (Extracts may remain discolored in the presence of ammonia even though no metal complex exists.)

7. Wash reagent. Dissolve 5 g. of potassium cyanide (*extremely poisonous!*) in 250 ml. of concentrated ammonium hydroxide and dilute to 500 ml. with water. This solution is stable at room temperature if stored in a borosilicate glass-stoppered bottle.

NOTE: Plastics other than Teflon will slowly react with this reagent and consequently should not be employed for its storage.

8. Stock standard, 1.0 mg. Pb/ml. Dissolve 0.183 g. of lead acetate trihydrate, $Pb(C_2H_3O_2)_2 \cdot 3H_2O$, in water containing 1 ml. of glacial acetic acid and dilute to exactly 100 ml. with water. This solution is stable at room temperature if stored in a tightly stoppered plastic bottle.

9. Working standards, 4.0 and 8.0 μg. Pb/ml. Prepare fresh standards daily by diluting the stock exactly 1:50 with water. Use aliquots of 2.00 and 4.00 ml. (corresponding to 4.0 and 8.0 μg. Pb/ml.) in step 2 of the procedure.

Procedure

1. Pipet into a glass-stoppered borosilicate 25 × 200 mm. test tube[2] 2.00 ml. of heparinized blood[3] or a 10.0 ml. aliquot of a urine sample.

NOTE: Blood may be collected with "B-D Vacutainer apparatus"[3] or with syringes. If syringes are used, they must be cleaned as described in the note above. Prevent the blood sample from clotting by mixing it with a small drop of heparin sodium solution (1 ml. = 1000 USP Units), added to the Vacutainer tube *immediately* after filling it with blood. If blood is obtained with a syringe, wet the syringe with the heparin just before the venipuncture.

NOTE: Some workers, particularly those affiliated primarily with industrial hygiene, use no anticoagulants and digest a given weight of "wet" clot (whole blood). This practice yields results only in terms of "μg. Pb per 100 g. 'wet' clot."

NOTE: Ideally a 24 hour urine sample is collected in a glass-stoppered borosilicate or polyethylene bottle containing 1–2 g. of EDTA (free acid) (2). If it is not convenient to obtain a 24 hour volume of urine, a single specimen may be collected in a bottle containing 100–200 mg. of EDTA (free acid) (2). In this case the results are expressed as micrograms Pb per liter of urine.

2. With each set of unknowns include tubes for a blank and two standards containing respectively 10.0 ml. of water and 2.00 and 4.00 ml. of working standard solution.

3. To all tubes add 5.0 ml. of digestion reagent and 3 borosilicate solid glass beads (approximately 5 mm. diameter).

4. In a fume hood heat each tube over a microburner until thin white fumes appear. Remove the burner and permit each tube to *cool to room temperature.* At this stage, blood is dark brown, urine samples are yellow, and the blank and standards are essentially colorless. Add 0.6 ml. of perchloric acid to each cool tube and cover with an inverted 50 ml. beaker.

NOTE: Checker (I. S.) recommends adding 1–2 ml. of a 1 + 1 mixture of HNO₃ and HClO₄.

5. Reheat each tube and continue heating while the solution goes through the following stages: yellow; colorless; yellow again with

[2] Corning Catalog No. 99650.

[3] Tubes for blood lead determination, Catalog No. L-3200, and disposable needles, Catalog No. H1004-20G, Becton, Dickinson and Co., Rutherford, New Jersey.

frothing; finally colorless and clear. Blood digests may remain very faintly yellow. At this point the digestion is complete. At times, particularly with urine, the solution turns colorless when the perchloric acid is added. It must, nevertheless, be heated through the yellow frothing phase to the colorless stage in order to complete the digestion. Allow the tube to cool to room temperature.

NOTE: Within a few minutes after addition of perchloric acid and application of heat, the standards and blank turn yellow, and after 20–30 minutes of additional digestion they are colorless and clear. The yellow urine samples are colorless immediately after addition of perchloric acid and heat. Within 2–4 minutes they are yellow again and after 25–35 minutes are colorless and clear. The dark brown blood digests fade slowly to yellow and in 40–60 minutes are colorless (R. B. F.)

NOTE: Checker (I. S.) adds 5.0 ml. of water to the cooled residue and repeats the digestion. This insures the elimination of all volatile oxides.

6. Add 5 ml. of water to each tube, followed by dropwise addition with mixing, 4 ml. of concentrated ammonium hydroxide. Allow the mixture to cool and add 1 drop of phenol red. If the solution has been neutralized sufficiently, it will remain red. If the proper alkalinity has not been attained, add additional ammonium hydroxide dropwise until the solution is red.

NOTE: A reddish color may appear after the addition of a small volume of ammonium hydroxide solution, and an inexperienced analyst may be misled into thinking that neutrality has been reached. This is potentially dangerous since the potassium cyanide solution might be added while the solution is still acid with the liberation of hydrogen cyanide. Moreover, the pH for the dithizone extraction would be incorrect. The color change of phenol red indicator goes from reddish to yellow to a final red. Jacobs and Herndon (4) prefer α-naphtholphthalein because this indicator is virtually colorless on the acid side and becomes green to blue-green without any intermediate indeterminate color on the alkaline side (pH 8). Thus, danger during the addition of cyanide is minimized.

NOTE: If the presence of bismuth is suspected in the sample (rare!), it can be removed by bringing the pH of the digestion solution to about 3.4 (indicator paper) with concentrated ammonium hydroxide and extracting with 5.0 ml. of dithizone solution. Discard the dithizone layer, using a capillary pipet and suction. Add the indicator, alkalinize with ammonium hydroxide, and continue according to step 6.

NOTE: Checker (I. S.) adds 1.0 ml. of 20% (w/v) aqueous hydroxylamine hydrochloride at this point to remove potential interference from administered iron compounds.

7. Add 10.0 ml. of buffer solution to each *cooled* tube followed within 1 minute by 3.00 ml. of dithizone solution.

8. Stopper all tubes with a glass stopper and shake vigorously 20 times.

9. Using a capillary pipet and a water-aspirator draw off most of the supernatant aqueous reagent.

NOTE: The critical point can be observed by watching for a ring of liquid which rises from the surface of the chloroform layer when there is no longer enough water over the chloroform phase to maintain an intact water layer. This ring of water rises several millimeters up the side of the tube and indicates plainly when sufficient aqueous phase has been removed.

NOTE: If the dithizone layer is colored mauve or red, add an additional 3.00 ml. of dithizone solution, and shake the mixture again. Continue with the rest of the procedure. Double the absorbance to obtain the correct value for lead in the specimen.

NOTE: Checker (R. B. F.) recommends the initial addition of 5.00 ml. of dithizone as being more appropriate than 3.00 ml.

10. Add 10.0 ml. of wash reagent, reshake until clear, and remove the supernatant layer by aspiration, again watching for the sign of sufficient removal of the aqueous layer.

11. Wet a 7 cm. Whatman No. 541 filter paper disc with chloroform and filter the final chloroform solution from step 10 into a cuvet. Stopper promptly.

NOTE: Checker (I. S.) recommends centrifuging for 3 minutes at high speed at this point.

12. Determine the absorbance of the blank, standards, and unknowns at 510 mμ against a blank of chloroform. Although the color is stable for at least 24 hours, obtain all readings promptly in order to minimize the effect of evaporation.

Calculations

BLOOD

$$\mu g.\ Pb/100\ ml. = \frac{(A_{unk}) - (A_{bk})}{(A_{4\ \mu g.\ std} - A_{bk}) + (A_{8\ \mu g.\ std} - A_{bk})} \times 12 \times \frac{100}{2}$$

URINE

$$\mu g.\ Pb/24\ hr. = \frac{(A_{unk}) - (A_{bk})}{(A_{4\ \mu g.\ std} - A_{bk}) + (A_{8\ \mu g.\ std} - A_{bk})} \times 12 \times \frac{ml./24\ hr.}{10}$$

NOTE: Using a 1 cm. cuvet, the absorbance-concentration curve is linear to about 10 μg. of lead corresponding in concentration at the suggested amounts of blood or urine taken to 500 μg. of lead per 100 ml. of blood or to 1000 μg. per liter of urine. The amount of dithizone used is sufficient to react with all urinary lead up to a concentration of 1500 μg. per liter. However, the absorbance at this level is unreadable. Such deeply colored solutions can be diluted appropriately with chloroform and read without error. Blanks are consistent for any set of reagents, and the total blank is equivalent to approximately 1 μg. of lead.

Discussion

Bessman and Layne (1) reported that duplicate determinations of 42 consecutive urine specimens showed a mean difference of 2.16 ± 1.99% from the mean of any pair. Twenty-six consecutive whole blood determinations showed a mean difference of 3.70 ± 2.04% from the mean of any pair. They also reported data for duplicate determinations of 26 consecutive specimens of blood and 41 urine specimens from patients with lead poisoning at various stages of therapy. The agreement between duplicate determinations on urine was 3.86 ± 2.61%, and between duplicate blood levels was 7.45 ± 4.20%. Checkers (R. B. F.) and (I. S.) found an average recovery of 99% in experiments in which varying amounts of lead (4, 6, 8 μg.) were added to samples of urine. The analysis of a single unknown requires about 2.5 hours, but by using multiple digestion units, as many as 10 samples can be analyzed simultaneously without greatly increasing the total time of analysis.

Interpretations (1, 3, 5, 6, 7, 8)

Lead is absorbed slowly and incompletely from the gastrointestinal tract. When inhaled it can also be absorbed from the respiratory tract. Consequently, lead poisoning is usually chronic because the metal is slowly absorbed, but is excreted even more slowly. Thus, in exposed individuals lead tends to accumulate in the body and is stored by tissues, especially bone. Symptoms of poisoning may follow such storage when, due to other diseases, the stored lead is released into the general circulation. For the diagnosis of chronic plumbism there are no infallible standard criteria, although the presence of low grade anemia, stippling of erythrocytes, increased δ-aminolevulinic acid and/or urinary coproporphyrin, and the presence of elevated lead concentration in the urine and blood are generally used. Pallor, colic, palsy, convulsions, and encephalopathy may suggest lead toxicity. In children, the characteristic density beneath the epiphyses of long bones as seen in radiographic films should serve as a warning that the child may have absorbed abnormal amounts of lead.

About 90% of the blood lead is found in the erythrocytes. Symptoms of lead poisoning are due to circulating lead and not to that in fixed deposits. Normally, the concentration of lead in the blood falls within the limits of 1–60 μg./100 ml., and values under 30–40 μg. are regarded as the most probable range of "normal" values. The critical blood level of lead usually associated with onset of severe symptoms is above 80 μg./100 ml. of blood. Values of 60–80 μg./100 ml. of blood are indicative of abnormal absorption of lead, but often not a degree of absorption capable of inducing symptoms. Subjects with this concentration should be watched carefully, and repeat determinations are desirable.

Lead is excreted from the body in both the feces and urine. The true alimentary excretion of lead is small, though most of the lead in feces comes from having been swallowed but not absorbed. Most of the *absorbed* lead is excreted by the kidneys. Hence the amount of urinary excretion of lead depends on both the lead exposure and on gastrointestinal absorption. The "normal" daily urinary excretion by individuals who have no known exposure to lead is 10–75 μg. and about 250 μg. per 24 hour fecal sample. The demonstration of 50 μg. or more of lead per liter of urine is generally a reasonable indication of lead poisoning. On the other hand, in people of no known exposure, up to 50 μg. per liter (about 75 μg. per day) may be found. The borderline seems to be about 80–100 μg. per day. As with blood, repeat determinations are desirable.

Administration of calcium EDTA to patients who have abnormal amounts of lead in their tissues is followed by a rise in urinary excretion of lead. This increased lead excretion following "edathamil" administration has been suggested as the basis for diagnostic test. Intoxicated patients, or those with increased lead absorption who generally receive edathamil therapy, excrete 1.0 mg. or more of lead in the urine on the first day of treatment. Normal patients subjected to this treatment excrete up to 700 μg. of lead in the first day's urine. This test may be more practical than the collection and analysis of a 24 hour urine specimen from an untreated patient in those instances where a delay in therapy is hazardous.

Excessive porphyrinuria (coproporphyrin III) is a constant finding in acute episodes of lead intoxication, and an early sign of chronic lead poisoning. Fifty per cent of subjects with 50 μg. of lead per 100 ml. of blood exhibit porphyrinuria; with higher lead concentrations, virtually all subjects excrete porphyrin.

REFERENCES

1. Bessman, S. P., and Layne, E. C., Jr., A rapid procedure for the determination of lead in blood or urine in the presence of organic chelating agents. *J. Lab. Clin. Med.* **45**, 159–166 (1955).
2. Keenan, R. G., Byers, D. H., Saltzman, B. E., and Hyslop, F. L., The "USPHS" method for determining lead in air and in biological materials. *Am. Ind. Hyg. Assoc. J.* **24**, 481–491 (1963).
3. Moncrieff, A. A., Koumides, O. P., Clayton, B. E., Patrick, A. D., Renwick, A. G. C., and Roberts, G. E., Lead poisoning in children. *Arch. Disease Childhood* **39**, 1–12 (1964).
4. Jacobs, M. B., and Herndon, J., Simplified one color dithizone method for lead in urine. *Am. Ind. Hyg. Assoc. J.* **22**, 372–376 (1961).
5. Stolman, A., and Stewart, C. P., Metallic poisons. In "Toxicology: Mechanisms and Analytical Methods" (C. P. Stewart and A. Stolman, eds.), Vol. 1, pp. 215–217. Academic Press, New York, 1960.
6. "Lead Poisoning in Children: Diagnostic Criteria." National Clearinghouse for Poison Control Centers, U.S. Dept. of Health, Education and Welfare, Washington, D.C., May, 1959.
7. Griggs, R. C., Sunshine, I., Newill, V. A., Newton, B. W., Buchanan, S., and Rasch, C. A., Environmental factors in childhood lead poisoning. *J. Am. Med. Assoc.* **87**, 703–707 (1964).
8. "Lead Poisoning in Children;" Editorial. *Lancet* Vol. 1, 867–868, 1964.

MAGNESIUM (FLUOROMETRIC) *

Submitted by: RALPH E. THIERS, Duke University Medical Center, Durham, North
  Carolina
Checked by: FRANK MADERA-ORSINI, St. Joseph Mercy Hospital, Ann Arbor,
  Michigan
  RICHARD W. MYERS, St. Francis Hospital, and SAMUEL BELFER,
  The Belfer Laboratories, Peoria, Illinois
  GEORGE H. WIEN, Morristown Memorial Hospital, Morristown,
  New Jersey

Introduction

Precipitation with 8-hydroxyquinoline has long been a preliminary step in the determination of magnesium by a variety of techniques (1, 2, 3). Schachter (1959) noted the fluorescence of the magnesium compound of 8-hydroxyquinoline (I) in ethanol and used this property for quantitative fluorometric assay (4). In 1961, he improved the method by employing the water-soluble reagent, 8-hydroxy-5-quinolinesulfonate (III), thus eliminating the need for alcoholic solutions (5).

Other methods employed for the determination of magnesium in serum or urine include gravimetric, colorimetric, flame photometric, and titrimetric procedures. The fluorometric procedure described here possesses distinct advantages in that it is specific, simple, accurate, rapid, and employs a very small sample.

Principle

The reagent, 8-hydroxy-5-quinolinesulfonic acid (III), forms fluorescent chelate compounds in aqueous solution with magnesium (II), calcium, zinc, and certain other heavy metals. Of these, only magnesium and calcium are present in serum in significant concentrations. The parent compound of this reagent, 8-hydroxyquinoline, has a high practical specificity for magnesium as compared with calcium (6). Both the metal-binding and molar fluorescence constants are much greater for the magnesium than the calcium complex.

* Based on the method of Schachter (5).

131

$$\underset{(I)}{\text{(structure I)}} \qquad \underset{(II)}{\text{(structure II)}} \qquad \underset{(III)}{\text{(structure III)}}$$

Because 8-hydroxy-5-quinolinesulfonic acid also has these properties, and, in contrast to the parent compound, is water soluble, it can be more readily used in a fluorometric method for serum magnesium. Both compounds I and II fluoresce, when irradiated with ultraviolet light, much more strongly than 8-hydroxyquinoline or its 5-sulfonate (III) presumably because of the changes in ring structure seen in formulas I and II as compared with III (6). The intensity of the increased fluorescence, which one measures in fluorometry, is directly proportional to the amount of magnesium present over a considerable concentrational range.

Reagents

1. Tris buffer, 2.0 M, pH 7.0, stock solution. Place 121 g. of Tris base, tris(hydroxymethyl)aminomethane, $NH_2C(CH_2OH)_3$, in 200 ml. of water. Adjust to pH 7.0 (pH meter) by adding concentrated HCl (12 N). About 75 ml. of the acid is required. Heat is produced in this neutralization: carefully cool the solution to room temperature before making the final pH adjustment. Dilute to 500 ml. with water. This reagent is stable.

2. 8-Hydroxy-5-quinolinesulfonic acid, 0.05 M, stock solution. Add 5.65 g. of 8-hydroxy-5-quinolinesulfonic acid (III)[1] to 27.5 ml. of 1.0 N NaOH and dilute to 500 ml. with water. Store in a brown bottle at 5°C. The solution is stable for at least 2 months.

3. Working buffer solution. To 1 vol. of Tris buffer solution add 19 vol. of water. This solution is stable.

4. Working reagent solution. To 1 vol. of stock sulfonic acid solution add 1 vol. of stock Tris buffer solution and 18 vol. of water. This solution is stable for only 2 or 3 days.

[1] The reagent marketed by Distillation Products Industries and Fisher Scientific Co. is satisfactory.

5. *Magnesium stock standard solution, 50.0 meq./l.* Dry reagent grade magnesium oxide at 800°C. for 4 hours. Dissolve 100.8 mg. of the dry powder in approximately 2 ml. of 12 N HCl and dilute to 100 ml. with water. This solution is stable.

6. *Working magnesium standards.* Make a series of working magnesium standards as follows:

Ml. stock standard diluted to 100 ml.	Concentration, meq./l.
1.00	0.50
2.00	1.00
3.00	1.50
4.00	2.00
5.00	2.50
6.00	3.00

These standard solutions are stable.

Procedure for Serum

Any fluorometer capable of excitation and detection in reasonably narrow band widths should be satisfactory for this method. Maximum fluorescence of the magnesium complex occurs at 510 mμ with an exciting wavelength of 405 mμ.[2]

Tubes were selected and prematched by the following procedure: Add 5.0 ml. of the working standard containing 2.50 meq./l. of Mg to 200 ml. of working reagent. Mix thoroughly and place 4.0 ml. of this solution into each of the tubes to be matched. Measure the total fluorescence of each tube and discard all those which vary from the mean by more than ±0.5 fluorescence units (the inherent precision of the instrument).

Careful cleaning of the tubes is essential. They may be soaked in 2 N HNO$_3$ and then in a 0.5% solution of the sodium salt of EDTA followed by thorough rinsing with deionized water. Another very effective technique is to rinse the tubes shortly before use with the

[2] Two different instruments were employed by the Submitter and Checkers: (a) G. K. Turner Associates (Palo Alto, California) Model 111, with 100–005 sample compartment, 110–850 lamp, 405 primary filter and 2A-12 secondary filter at 10× sensitivity. (b) Zeiss PMQ II (Carl Zeiss, N.Y.) with fluorescence attachment, 405 primary filter, slit width 0.5 mm, detector wavelength 510 mμ, full sensitivity.

134

134

working reagent solution (since it binds magnesium tightly) and allow them to drain dry.

1. Adjust the fluorometer scale to 0 with a tube containing 4.0 ml. of working buffer solution.

NOTE: All subsequent determinations are made in duplicate and fluorometer scale readings are averaged. Each specimen analyzed requires four tubes, two of which serve as serum blanks.

2. Read two or more Reagent Blank (RB) tubes containing 4.0 ml. of working reagent solution. Take the average of the readings. This need be done only once daily.

3. Read two unknown (R) tubes containing 4.00 ml. of working reagent and 0.100 ml. of serum.

4. Read two Serum Blank (SB) tubes containing 4.00 ml. of the working buffer and 0.100 ml. of serum.

5. Read two tubes containing 4.00 ml. of working reagent and 0.100 ml. of 2.00 meq./l. magnesium working standard. This need be done only once with each batch of analyses, and no blank is needed for these standards.

NOTE: If the temperature inside the fluorometer tends to warm the tubes of solution appreciably the readings should be taken quickly, because fluorescence intensity decreases markedly with increasing temperature—about 2% per °C.

Calculations

1. Obtain the net fluorescence (NF) of the serum by subtracting the sum of the average readings of the Reagent Blank and the Serum Blank from the average readings of the unknowns:

$$NF = R - (RB + SB)$$

2. Obtain the NF for the standard:

$$NF = R - RB$$

3. Calculate:

$$\text{meq./l. of Mg in serum} = \frac{NF \ (\text{serum})}{NF \ (\text{standard})} \times 2.00$$

The linearity of the relationship between fluorescence and magnesium concentration should be determined by use of the working standard solutions whenever new reagents are introduced. This relationship is linear up to at least 3 meq./l. (and if the sensitivity of the fluorometer is set correctly, higher concentrations will give readings off-scale). More concentrated samples should be diluted prior to analysis (and checked for possible contamination).

Discussion

Schachter tested a number of divalent cations (including Ca) and only Cd, Zn, Sn, Sr, and Ba showed appreciable fluorescence under the conditions of this method. As these are present in negligible quantities in serum and urine the method is specific for magnesium.

The fluorometric determination of magnesium offers several advantages over other methods. It does not require the precipitation necessary, for example, in the Titan Yellow method, and is more sensitive and specific than either the Titan Yellow and flame photometric methods. It possesses potential disadvantages in that fluorescence can be enhanced or depressed by certain substances; however, these problems appear to arise extremely rarely with serum or plasma.

The precision of the method was determined by taking the variation between duplicate or triplicate data on over sixty different samples. The standard deviation obtained was 0.03 meq./l., and the coefficient of variation was 2%.

Recovery experiments and analyses of commercial standard samples have indicated that the method is accurate to within the reproducibility of the measurements.

There has been some disagreement about the range of magnesium concentrations in normal human serum. This has been attributed to difficulty in the preparation of accurate standards. As a standard the Submitter used a solution which was gravimetrically analyzed independently by two different analysts to assure correspondence between the actual and expected concentration. The Submitter and Checkers made independent determinations of magnesium in serum samples from apparently healthy persons, mixed with respect to age and sex. The data obtained, along with that of the clinical biochemistry laboratory at the University of Michigan Hospital, are shown in Table I.

Although Schachter recommends this method for the determination of magnesium in urine, Stewart and Frazer (7) state "In urine . . . we have found frequent and severe fluorescence quenching by unidentified constituents, in some cases so marked as to reduce the apparent recoveries of added magnesium to between 25 and 40%." Presumably, some separation step or ashing of the urine would overcome this problem.

TABLE I

NORMAL VALUES FOR SERUM MAGNESIUM

Laboratory	Number of persons[a]	Mean	Range
Duke University Medical Center	58	1.89	1.6–2.2
Morristown Memorial Hospital	27	1.61	1.4–1.8
St. Joseph Mercy Hospital	30	1.85	1.3–2.4
University of Michigan Hospital	60	1.88	1.4–2.3

[a] Adults of both sexes.

REFERENCES

1. Hahn, F., Fortschritte in der analytischen Chemie der Leichtmetalle. Z. angew. Chem. 39, 1198 (1926).
2. Hoffman, W. S., A colorimetric method for the determination of serum magnesium based on the hydroxyquinoline precipitation. J. Biol. Chem. 118, 37–45 (1937).
3. Davis, S., A flame photometric method for the determination of plasma magnesium after hydroxyquinoline precipitation. J. Biol. Chem. 216, 643–651 (1955).
4. Schachter, D., The fluorometric estimation of magnesium in serum and in urine. J. Lab. Clin. Med. 54, 763–768 (1959).
5. Schachter, D., Fluorometric estimation of magnesium with 8-hydroxy-5-quinolinesulfonate. J. Lab. Clin. Med. 58, 495–498 (1961).
6. Watanabe, S., Frantz, W., and Trottier, D., Fluorescence of magnesium-calcium-, and zinc-8-quinolinol complexes. Anal. Biochem. 5, 345–359 (1963).
7. Stewart, C. P., and Frazer, S. C., Magnesium. Advan. Clin. Chem. 6, 29–65 (1963).

MAGNESIUM (TITAN YELLOW)*

Submitted by: Daniel H. Basinski, Henry Ford Hospital, Department of Laboratories, Detroit, Michigan

Checked by: James L. Gilleland, Providence Hospital, Waco, Texas

George W. Johnston, Chemistry Section, Third United States Army Medical Laboratory, Fort Mcpherson, Georgia

Kenneth M. Takehara, Conemaugh Valley Memorial Hospital, Johnstown, Pennsylvania

Norbert W. Tietz, Division of Biochemistry, Mount Sinai Hospital, Chicago, Illinois

Introduction

Magnesium levels in serum are normally quite stable. However, in certain pathological states such as chronic and acute renal failure, the magnesium level is sometimes elevated (1), whereas in alcoholic cirrhosis and after prolonged parenteral fluid administration it may be decreased (2). Older methods involving precipitation of magnesium either as magnesium ammonium phosphate (3) or as molybdivanadate (4) are capable of satisfactory accuracy but have the disadvantage of being time-consuming and cumbersome. Flame photometry, for best results, requires multichannel instruments (5) not readily available in most clinical laboratories. Atomic absorption also yields excellent results (6) but again requires equipment not yet in wide use. Colorimetric procedures such as that of Orange and Rhein (7), while rapid and simple, yield erratic results. The method presented here is the result of a systematic study of the factors which contribute to variability, namely, alkali concentration, dye concentration, and calcium interference. The use of a recording spectrophotometer facilitated the working out of the interactions of reagent concentrations and changes in peak absorbance. It was thus possible to choose a set of optimal conditions for a rapid, simple, and reproducible method.

Principle

The magnesium in a trichloroacetic acid filtrate of serum forms a red-colored lake in alkaline solution with the dye, Titan Yellow.

* Based on modification of the method of Orange and Rhein (7).

137

138   DANIEL H. BASINSKI

The lake is stabilized with polyvinyl alcohol and its absorbance measured in a spectrophotometer.

Reagents

1. Trichloroacetic acid, 5.0% (w/v). Dissolve 50.0 g. of trichloroacetic acid in distilled water and make up to 1 l. Store the solution in a glass-stoppered borosilicate bottle. The normality should be 0.306 when checked with standard sodium hydroxide.

NOTE: Checker (N. W. T.) recommends the use of deionized water in this method. The Submitter has not found this necessary. However, if any doubt exists as to the purity of the distilled water used, this precaution should be observed.

2. Sodium hydroxide, 2.50 N. Dilute 100 ml. of a 50% (w/v) solution of sodium hydroxide to 500 ml. with boiled and cooled distilled water. Check the normality by titration with standard acid. Store in a borosilicate bottle securely stoppered with a rubber or polyethylene stopper.

3. Polyvinyl alcohol, 0.1% (w/v). Suspend 1.0 g. of polyvinyl alcohol[1] (PVA) in 40–50 ml. of 95% ethanol and pour the mixture into 500–600 ml. of swirling distilled water. Warm the solution on a hot plate until it is clear. Allow the solution to cool to room temperature and dilute to 1 l.

NOTE: Polyvinyl alcohol can be dissolved directly in water, but it quickly forms lumps which take a long time to dissolve completely. Dispersion in alcohol prevents lumping.

4. (a) Titan Yellow stock solution. Dissolve 75 mg. of Titan Yellow in 0.1% PVA and make up to 100 ml. Store in a brown bottle at room temperature.

(b) Titan Yellow working solution. Dilute the stock solution 1:10 with 0.1% PVA.

NOTE: The stock Titan Yellow solution may be filtered if it is not clear. The working solution will give reliable results for a week if stored in a brown bottle at room temperature. The stock solution will keep for at least 2 months. Titan Yellow is also known as Clayton Yellow and as Thiazole Yellow.[2]

[1] The polyvinyl alcohol used in the Submitter's laboratory was obtained from DuPont under the label Elvanol Grade 70-05. Checker (K. M. T.) purchased PVA from K & K Laboratories, Inc., Plainview, New York, and reported that it worked as well as Elvanol.

[2] Eastman Kodak Company, Distillation Products Industries, Titan Yellow, Cat. No. P 4454. Clayton Yellow, Cat. No. 1770.

5. *Standard magnesium solution, 20 meq./l.* Dissolve 243.2 mg. of bright magnesium metal turnings (reagent grade, suitable for Grignard's reaction) in a covered beaker with 50 ml. of distilled water and 2 ml. of concentrated hydrochloric acid. Avoid open flames as hydrogen is evolved. After the vigorous reaction subsides, add sufficient hydrochloric acid dropwise to completely dissolve the magnesium metal. Transfer quantitatively to a 1 l. volumetric flask and make up to volume with distilled water. Store in a tightly stoppered borosilicate bottle. This solution will keep indefinitely.

6. *Working standard magnesium solutions.* Accurately pipet 5.00, 10.00, 15.00, and 20.00 ml. aliquots of stock standard magnesium (20 meq./l.) into 100 ml. volumetric flasks and dilute with water to exactly 100 ml. The standards contain 1.0, 2.0, 3.0, and 4.0 meq. magnesium per liter. Store in tightly stoppered borosilicate or polyethylene bottles.

Procedure

1. Pipet 5.00 ml. of 5% trichloroacetic acid into a 15 ml. test tube and slowly, with mixing, add 1.00 ml. of serum.

2. Mix gently but thoroughly and centrifuge for 5 minutes at 2,000 r.p.m.

3. Transfer 3.00 ml. of the clear supernatant to a 19 × 105 mm. round cuvet or test tube, add 2.00 ml. of the Titan Yellow working solution, and 1.00 ml. of 2.50 N NaOH. Mix thoroughly.

NOTE: Turbidity in the trichloroacetic acid supernatant fluid sometimes occurs but disappears on addition of alkali. Apparently it does not interfere with the determination.

4. Prepare a series of standards by pipetting 0.50 ml. of magnesium standard solutions into 19 × 105 mm. cuvets or test tubes, adding 2.50 ml. of 5% trichloroacetic acid, followed by 2.00 ml. of the Titan Yellow and 1.00 ml. of the 2.50 N NaOH. A reagent blank is prepared at the same time, substituting distilled water for the magnesium standard solution.

5. The absorbances of samples and standards are read at 540 mμ in a spectrophotometer with the reagent blank set at zero absorbance.

NOTE: The reaction appears to be instantaneous and readings may be made without delay. The colored lake which forms is stable within reasonable limits. However, slight changes in absorbance will take place over a prolonged period of time so that the best results will be obtained if the solutions are read within one-half hour.

The absorption peak for this determination is quite critical. The spectrophotometer used should be carefully calibrated by means of a didymium glass filter and the maximum absorbance of the reaction mixture should be determined for that particular instrument. The Checkers reported a variation in the peak absorbance ranging from 530 to 540 mμ.[3] Each experimenter must determine the maximum absorbance wavelength for his instrument. Because of the sharp absorbance peak filter photometers are not suitable for this determination.

Calculation

Magnesium values are calculated from the formula:

$$\text{meq. Mg/l.} = \frac{A_{unk} \times C_{std}}{A_{std}}$$

where A_{unk} and A_{std} are absorbances of unknown and standard respectively, and C_{std} is the concentration of the standard magnesium solution most nearly corresponding to the value of the unknown. Ordinarily it is sufficient to use only the 1 and 2 meq. standards.

NOTE: With narrow-band spectrophotometers the color reaction adheres to Beer's law. Relatively broad-band instruments, such as the Coleman Jr. spectrophotometer, show some deviation from a straight line at higher concentrations.

Discussion

Inasmuch as the level of magnesium in red blood cells is approximately three times higher than in serum, visible hemolysis is cause for rejection of a specimen. Another source of error is the interference caused by recent administration of calcium gluconate (8). Any abnormally low values should be regarded with suspicion. Reference to the patient's hospital record may show that calcium gluconate was involved, in which case the value of serum magnesium by this method is unreliable, and the clinician should be so warned.

Under the conditions of this procedure there is no interference by calcium, nor has the Submitter found any evidence of change in magnesium values on storage of sera, either at 4°C. or frozen. Precipitation of protein by trichloroacetic acid in this method releases magnesium quantitatively.

In the course of the development of this modification, a series of 97

---

[3] Repeated determinations confirmed the validity of 540 mμ for the Coleman Jr. spectrophotometer used in the Submitter's laboratory.

specimens was run for which the values of sodium and potassium were known to be normal. The mean value for magnesium was 1.80 meq./l., with a standard deviation of 0.26 meq./l. The two-sigma range for this presumably normal series is therefore 1.3 to 2.3 meq./l. Analysis of 331 "unknown" sera showed a mean of 1.72 meq./l., with a standard deviation of 0.31. An analysis of 38 random serum samples by Checker (G. W. J.) yielded a mean value of 1.77 meq. He uses a normal range of 1.3 to 2.0 meq./l. Checker (J. L. G.) collected a 2 l. pool of serum over a period of several months. Replicate analyses have shown a mean concentration of 1.92 meq. of magnesium per liter.

The Submitter's experience of normal mean value of 1.80 meq./l. for serum magnesium seems to be a reasonable compromise. In a review article, MacIntyre (9) claims a measure of agreement for a value of 1.66 meq/l., three references being quoted, two of the results attained by flame spectrometry and the third by a precipitation method. By contrast, Wacker and Vallee (5), using a multichannel flame spectrometer, found a mean of 2.05 meq./l. for 14 normal subjects. Elkinton (10) has tabulated groups of values from the literature. Nine references for precipitation method mean values ranged from 1.66 to 2.07 meq./l., while five references in which Titan Yellow procedures were used showed mean values from 1.76 to 1.91 meq./l. Such a relatively broad spread of mean values determined by precipitation methods and by flame spectrometry seems to imply that all methods appear to be equally good—or equally bad. An argument in favor of the simple colorimetric Titan Yellow procedure can be found in the report by Gorfien and Kramer (6) in which atomic absorption values closely check those obtained by a modified Titan Yellow method. Those laboratories lacking the more sophisticated equipment can therefore report equally valid results derived by colorimetry.

REFERENCES

1. Hamburger, J., Electrolyte disturbances in acute uremia. *Clin. Chem.* **3,** 332–343 (1957).
2. Flink, E. B., Stutzman, F. L., Anderson, A. R., Konig, T., and Fraser, F., Magnesium deficiency after prolonged parenteral fluid administration and after chronic alcoholism complicated by delirium tremens. *J. Lab. Clin. Med.* **43,** 169–183 (1954).
3. Kramer, B., and Tisdall, F. F., A simple technique for the determination of calcium and magnesium in small amounts of serum. *J. Biol. Chem.* **47,** 475–481 (1921).

4. Simonsen, D. G., Westover, L. M., and Wertman, M., The determination of serum magnesium by the molybdivanadate method for phosphate. *J. Biol. Chem.* **169**, 39–47 (1947).

5. Wacker, W. E. C., and Vallee, B. L., A study of magnesium metabolism in acute renal failure employing a multichannel flame spectrometer. *New Engl. J. Med.* **257**, 1254–1262 (1957).

6. Gorfien, P. C., and Kramer, B., Quantitative determination of magnesium in biological fluids: a comparative study. *Federation Proc.* **22**, 457 (1963).

7. Orange, M., and Rhein, H. C., Microestimation of magnesium in body fluids. *J. Biol. Chem.* **189**, 379–386 (1951).

8. Anast, C. S., The unreliability of the Titan Yellow method for the determination of magnesium in patients receiving intravenous calcium gluconate. *Clin. Chem.* **9**, 544–551 (1963).

9. MacIntyre, I., and Wootton, I. D. P., Clinical biochemistry. II. Magnesium metabolism. *Ann. Rev. Biochem.* **29**, 642–648 (1960).

10. Elkinton, J. R., The role of magnesium in the body fluids. *Clin. Chem.* **3**, 319–331 (1957).

Pertinent Review

Wacker, W. E. C., and Vallee, B. L., Magnesium metabolism. *New Engl. J. Med.* **259**, 431–438, 475–482 (1958).

METHEMOGLOBIN*

Submitted by: ADRIAN HAINLINE, JR., Duke University Medical Center, Durham, North Carolina†
Checked by: JOSEPH H. BOUTWELL, JR., and WANDA WILKES, Temple University School of Medicine and Hospital, Philadelphia, Pennsylvania
GEORGE N. BOWERS, JR., and JEAN WENZEL, Hartford Hospital, Hartford, Connecticut
W. L. FREAS, JR., St. Elizabeth Hospital, Dayton, Ohio
JACOB KREAM and DONATO BROGNA, Hospital for Joint Diseases, New York, New York
S. A. MORRELL and DAN WALDSCHMIDT, Milwaukee Blood Center, Milwaukee, Wisconsin
HENRY SHARTON, Aultman Hospital, Canton, Ohio

Introduction

The distinctive brown color of methemoglobin makes possible its spectrophotometric detection and estimation in the presence of normal hemoglobin pigments. Formerly, the hand spectroscope was used to observe the visible absorption spectrum of methemoglobin and other hemoglobin derivatives. Since the results obtained with a hand spectroscope were only qualitative (1), it was necessary for early investigators to use oxygen capacity methods for the indirect measurement of methemoglobin in blood (2). Quantitative spectrophotometric methods (3, 4, 5) were developed before the spectrophotometer was widely adopted in the clinical laboratory. Hamblin and Mangelsdorff (6) and, more recently, Hutchinson (7) described variations of the older spectrophotometric methods that are designed for use with the instruments found in the usual modern laboratory.

The method presented here adds an important modification to the work of Evelyn and Malloy (5). This modification is the use of a Triton-borate solution that acts as a hemolyzing and clearing agent in the diluent to remove the turbidity caused by the stroma of red cells, thus eliminating the need for prolonged centrifugation (8, 9).

Principle

The spectrophotometric determination is based on the measurement of absorbance of a diluted specimen at a wavelength where the

* Based on a modification of the method of Evelyn and Malloy (5).
† Present address: St. Luke's Hospital, Kansas City, Missouri.

143

absorbance of methemoglobin is distinct from that of oxyhemoglobin and hemoglobin. This wavelength is not the point of the maximum absorbance of methemoglobin (630 mμ) but the wavelength (620 mμ) where the absorbance of methemoglobin does not vary with pH (isobestic point) (4, 6). Thus, buffers are not needed (see Discussion).

After the absorbance is read at 620 mμ, cyanide is added to convert methemoglobin to cyanmethemoglobin. Cyanmethemoglobin and oxyhemoglobin absorb poorly at 620 mμ. The absorbance is read again and subtracted from the first reading to give a value that is proportional to the concentration of methemoglobin.

Total hemoglobin is determined as cyanmethemoglobin (10) in order that the methemoglobin results may be expressed in terms both of absolute concentration and of percentage of the total hemoglobin.

Reagents

1. Ferricyanide-cyanide. Dissolve the following reagents in about 200 ml. of water in a 1 l. volumetric flask: 1 g. sodium bicarbonate, NaHCO$_3$; 0.05 g. potassium cyanide, KCN; 0.2 g. potassium ferricyanide, K$_3$Fe(CN)$_6$. Dilute to 1 l. and mix. Transfer to an amber bottle. As this solution is somewhat unstable, store for no longer than 1 month after preparation.

2. Cyanide, 10% (w/v). Dissolve 5 g. of either reagent grade potassium cyanide, KCN, or sodium cyanide NaCN (*Caution! Poisonous!*), in 35 ml. of water in a 50 ml. volumetric flask Dilute to volume and mix. Store in a polyethylene bottle. Discard when the solution becomes colored.

3. Potassium ferricyanide, 10% (w/v). In a 50 ml. volumetric flask or glass-stoppered graduated cylinder, place 5 g. of K$_3$Fe(CN)$_6$, reagent grade crystals. Add water to dissolve the crystals and dilute to the mark.

4. Triton-borate solution. Place 2.55 g. of sodium tetraborate Na$_2$B$_4$O$_7$·10H$_2$O, in a 100 ml. glass-stoppered graduated cylinder. Dissolve in sufficient water to make 67 ml. of solution. Add 33 ml. of Triton-X-100[1] and mix. Store in an alkali-resistant container.

5. Water, distilled. Use distilled or demineralized water not exceeding the pH range of 5.0 to 7.5 for the dilution of blood specimens. Hemoglobin in solutions outside this pH range tends to form hematin.

[1] Triton-X-100 may be obtained from Hartman-Leddon Co., 60th and Woodland Ave., Philadelphia 43, Pennsylvania.

Procedure

Before drawing blood, prepare tubes to receive the specimens according to the following directions. To *tube A* add 9.80 ml. of distilled water and 0.10 ml. of Triton-borate solution and mix.

NOTE: Use a pipet calibrated for rinsing (TC), to add the Triton-borate solution, because of its high viscosity.

In *tube B* place 5.00 ml. of ferricyanide-cyanide in preparation for a total hemoglobin estimation. Immediately after collection of blood by finger- or veni-puncture, transfer with a micropipet 0.10 ml. of blood to *tube A*, mix, and centrifuge. With another micropipet transfer 0.02 ml. of blood to *tube B*, mix, and allow to stand 10 minutes.

NOTE: In the intact red blood cell, methemoglobin is converted to hemoglobin by methemoglobin-reductase. Therefore, methemoglobin will slowly disappear from undiluted specimens. Analysis of undiluted specimens should begin within 1 hour of collection; otherwise, slightly elevated levels of methemoglobin might not be detected.[2]

Transfer the crystal-clear supernatant fluid from *tube A* to a cuvet. Obtain absorbance reading A_1 with water as a blank, at 620 mμ, in a spectrophotometer. Add 0.05 ml. of 10% cyanide[3] to the diluted specimen; mix, and obtain reading A_2 at 620 mμ.

Read *tube B* at 540 mμ in a photometer calibrated for total hemoglobin by the cyanmethemoglobin method (10, 11, 12) (see the section on Standardization).

Calculation

$$A_1 - A_2 = \Delta_{absorbance}$$

$$\frac{\Delta_{absorbance}}{K} = \text{g. Met Hb}/100 \text{ ml.}$$

The constant K is obtained with known amounts of hemoglobin (see sample calculation under Standardization).

$$\frac{A_1 - A_2}{K} = \text{g. Met Hb}/100 \text{ ml.}$$

[2] As a precaution against accidental loss of the specimen, it may prove desirable, if the collection is made by venipuncture, to place from 1 to 2 ml. of extra blood in a tube containing Versenate, oxalate, or heparin (no fluoride). Should it be necessary to repeat the analysis or make further studies, the specimen would be readily available, providing that too much time has not elapsed and the methemoglobin has not disappeared.

[3] The analyst may prefer to use 2–4 mg. of the granular form of potassium or sodium cyanide.

The number of grams of methemoglobin thus obtained may be reported in terms of percentage methemoglobin in total hemoglobin by use of the following equation:

$$\% \text{ Met Hb in total} = \frac{\text{g. Met Hb}/100 \text{ ml.}}{\text{g. total Hb}/100 \text{ ml.}} \times 100$$

(See the section on Standardization.) If the addition of cyanide does not produce a reading that approximates the absorbance at 620 mμ for a similar concentration of oxyhemoglobin, the specimen contains other abnormal pigments such as sulfhemoglobin.

NOTE: Extremely high levels of lipid might cause a turbidity that would also increase the reading at 620 mμ. Any turbidity invalidates the results.

The total hemoglobin determined in *tube B* may be read from a standard curve based on known concentrations of hemoglobin converted to cyanmethemoglobin (10, 12).

Standardization

The standardization of total hemoglobin should be made with specimens of blood taken from normal subjects and analyzed for total iron (10, 12). It also is acceptable to standardize the total hemoglobin method against commercial standards of cyanmethemoglobin prepared according to the specifications of the National Research Council (11).

If the cyanmethemoglobin method is not routinely used for hemoglobin, it may be preferable to obtain the absorbance of the cyanmethemoglobin from *tube B* in a 1 cm. cuvet at 540 mμ in a narrow-band-width spectrophotometer for which the millimolar extinction coefficient E of cyanmethemoglobin has been determined with known standards. Then,

$$\text{total hemoglobin g.}/100 \text{ ml.} = \frac{A \times 250 \times 1.652}{E}$$

where 250 represents the dilution and 1.652 is the grams of hemoglobin per milligram atom of iron per 100 ml. (11). By using this equation, the extinction coefficient for a specific instrument can be determined with a solution of known concentration of hemoglobin in the form of cyanmethemoglobin. [In the United States, the value 11.5 for the millimolar extinction coefficient, E has been accepted by the National Research Council (11). This value has been questioned by Zijlstra and VanKampen (13).]

To standardize the methemoglobin procedure, obtain a 2 to 3 ml. blood specimen [using potassium Versenate, potassium oxalate, or heparin as an anticoagulant, but not fluoride, which forms fluoro-methemoglobin (14)] from a normal person (a nonsmoker who has not been taking drugs of any kind for at least 1 week). Transfer 1.0 ml. of the well mixed whole blood to 9.0 ml. of water in a small container such as a 25 ml. Erlenmeyer flask. Add 0.10 ml. of Triton-borate solution, mix well, and centrifuge at 1000 g for 5 minutes. Transfer about 5.0 ml. of the clear supernatant fluid to a test tube. Prepare additional dilutions of the hemolysate in test tubes according to the following tabulation (15):

		Tube				
	Blank	1	2	3	4	5
Water (ml.)	9.90	9.70	9.50	9.40	9.30	9.10
Hemolysate (ml.)	0	0.20	0.40	0.50	0.60	0.80
Percentage of total	0	20	40	50	60	80

Add 0.10 ml. of 10% potassium ferricyanide to each tube and mix. Let stand 10 minutes to develop fully the methemoglobin. Read absorbance A_1 at 620 mμ for each dilution. Add 0.05 ml. of 10% potassium cyanide. Mix, and let stand for 2 minutes. Read absorbance A_2 at 620 mμ. The contents of *tube* 3 should be read also at 540 mμ to determine the total hemoglobin (1:200 dilution) as cyanmethemo-globin. With this last reading, calculate the grams of hemoglobin/100 ml. of *tube* 3, and the concentration of hemoglobin which each of the other tubes represents.

Standardization[4]

	Tube				
	1	2	3	4	5
Grams Met Hb/100 ml.	3	6	7.5[a]	9	12
A_1	0.070	0.140	0.175	0.210	0.280
A_2	0.025	0.040	0.045	0.055	0.075
$A_1 - A_2$	0.045	0.100	0.130	0.155	0.205
$(A_1 - A_2)/g.$	0.015	0.017	0.017	0.017	0.017

$$(A_1 - A_2)/g. = K = 0.017$$

[4] Sample calculation (from data obtained with the Beckman "B" spectro-photometer using a 1 cm. cuvet).

[a] Total hemoglobin determined by cyanmethemoglobin method after addition of cyanide.

ADRIAN HAINLINE, JR.

Unknown

$$A_1 = 0.100, \qquad A_2 = 0.030.$$
g. Met Hb/100 ml. $= (A_1 - A_2)/K = 0.070/0.017 = 4.1.$
$$\text{Total Hb} = 14.0.$$
$$\% \text{ Met Hb} = (4.1 \times 100)/14 = 29.$$

The data obtained by various laboratories may not be as ideal as the example supplied by the Submitter. However, constants for high concentrations of methemoglobin obtained with 19-mm. cuvets in Coleman Jr. spectrophotometers used by the Submitter and Checkers agreed well: 0.019 ± 0.02. With the instruments used, all investigators noted that the constant for 3 g./100 ml. methemoglobin or less was lower than constants determined at higher concentrations. For a greater accuracy at lower concentrations, use a constant determined for the specific range desired.

Discussion

The spectrophotometric method for the determination of methemoglobin is based on differences in light absorption of the various hemoglobin pigments. In the individual who has neither been exposed to drugs nor to a contaminated atmosphere, the circulating hemoglobin is principally in the form of oxyhemoglobin and hemoglobin (reduced form). However, small amounts of methemoglobin are present in the normal subject (14, 16). Blood of smokers and those exposed to the exhausts of combustion engines also contains some carboxyhemoglobin (carbon monoxide hemoglobin).

All of these hemoglobin derivatives absorb light strongly in the range of 410–420 mμ. Differences in their absorbance peaks in this spectral range are insufficient for analytical purposes. However, in the range between 460–700 mμ the absorbances of the various pigments show characteristic differences (1, 17). Oxyhemoglobin has absorption maxima at 540 and 578 mμ[5]; carboxyhemoglobin has absorption maxima at 538 and 570 mμ[5] (Fig. 1). The location of oxyhemoglobin and methemoglobin absorption maxima differ greatly (Fig. 2). Note the absorption peak of methemoglobin at 635 mμ,[5] which is in a range where both oxyhemoglobin and carboxyhemoglobin have little absorbance.

The situation is somewhat complicated by the fact that methe-

[5] Various authors report slightly different absorption maxima (and minima) for the various hemoglobin derivatives. It should be emphasized that these variations are probably due to discrepancies in the calibrations of the instruments used. Therefore, accurate work, particularly with narrow-band spectrophotometers, will require that these points be checked on the instrument to be used.

moglobin absorption is affected by variations in pH (Fig. 3). The absorption maximum at 635 mμ is most pronounced in the acid form, and practically disappears in the alkaline form of methemoglobin. However, the absorbances at 620 mμ for the same concentrations do not change with pH (isobestic point). Another isobestic point for

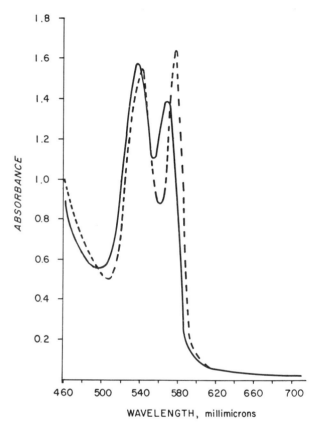

FIG. 1. Absorption spectra of equal quantities of hemoglobin as oxyhemoglobin (_ _ _ _ _), and as carboxyhemoglobin (_____) in 0.007 N NH₄OH.

methemoglobin is 520 mμ. This point was utilized by Hutchinson (7) as a means of measuring total hemoglobin, because methemoglobin absorbance is equal to that of oxyhemoglobin at 520 mμ.

Methemoglobin is converted by cyanide to cyanmethemoglobin, a stable pigment with a broad-band absorbance at 540 mμ and low

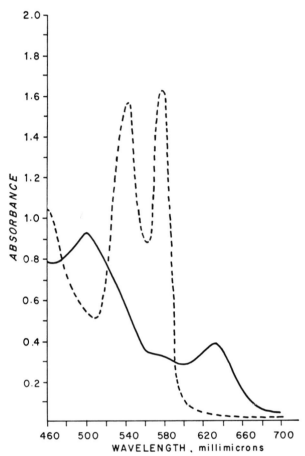

FIG. 2. Absorption spectra of equal quantities of hemoglobin as oxyhemoglobin
(_ _ _ _ _) and as methemoglobin (_____). Methemoglobin produced by
diluting specimen in 0.1% $K_3Fe(CN)_6$ in $M/60$ phosphate buffer, pH 6.0.

absorbance above 620 mμ (Fig. 4). The addition of cyanide to a solution containing methemoglobin will cause a decrease in the absorbance
in the range of the 635 mμ maximum, in proportion to the methemoglobin concentration. Since the absorbance of methemoglobin at
620 mμ is independent of pH, measurements of absorbances at that
wavelength eliminate the necessity of pH control by the use of buffers
that would be required if the measurements were made at 635 mμ.

Hutchinson (7) estimated methemoglobin from the ratio of the absorbances at the isobestic points 620 mμ and 520 mμ. The absorbance at 620 mμ is proportional to the methemoglobin concentration, and the absorbance at 520 mμ is proportional to the total hemoglobin. However, if sulfhemoglobin is present, the increase in absorbance at

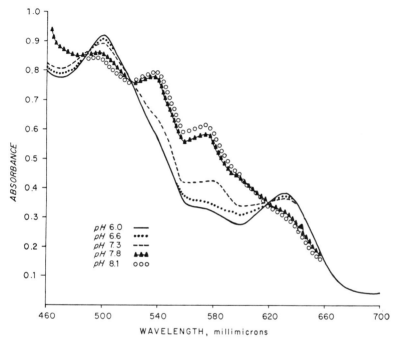

Fig. 3. Absorption spectra of equal quantities of methemoglobin in 0.1% K$_3$Fe(CN)$_6$ in M/60 phosphate buffers to demonstrate the variation of absorbance with pH.

620 mμ will be disproportionate to the methemoglobin concentration because the absorptivity coefficient of sulfhemoglobin is greater than that of methemoglobin at this wavelength. Sulfhemoglobin has an absorption maximum at 620 mμ (18) and does not react with cyanide (Fig. 5). Unless cyanide addition is included in the procedure, the presence of sulfhemoglobin would not be detected, and a considerable error would be introduced in the measurement of methemoglobin (see p. 155 concerning methemoglobin M.)

152 ADRIAN HAINLINE, JR.

The method described includes the addition of cyanide, and is based on the premise that any decrease in absorbance at 620 mμ after the addition of cyanide is due to the conversion of methemoglobin to cyanmethemoglobin. Another possible source of error in any measurement made on a hemolysate is the turbidity caused by the cellular remnants of red blood cells. These are difficult to remove by centrifugation, and unfortunately tend to dissolve upon the addition of cyanide. However, this turbidity can be prevented or minimized

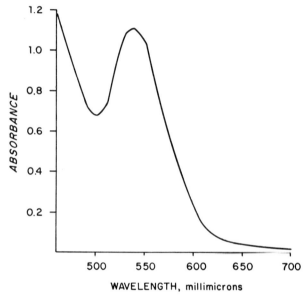

Fig. 4. Absorption spectrum of cyanmethemoglobin in ferricyanide-cyanide solution.

by the use of surface-acting agents such as Triton-X-100 (9). Care is necessary to note whether or not turbidity is obviously different after the addition of cyanide.

NOTE: The use of a narrow-band-width spectrophotometer is necessary if a high sensitivity is desired. In most clinical situations such sensitivity is not required, and the use of wide-band-width spectrophotometers appears to suffice. It must be emphasized that replication of wavelength settings is extremely critical, especially when one is using a narrow-band-width instrument. By utilizing the information available from the absorption curves of the various hemoglobin pigments it is also possible to determine total hemoglobin, oxyhemoglobin, oxyhemoglobin/reduced hemoglobin ratios, and carboxyhemoglobin. All such methods require exact reproduction of the wavelength settings used in calibration and measurement.

The estimation of methemoglobin may also utilize the differences in absorbances in the near infrared, such as has been reported by Horecker and Brackett (19). As has been mentioned, early measurements were made with gasometric techniques, which are in general more difficult than the spectrophotometric methods.

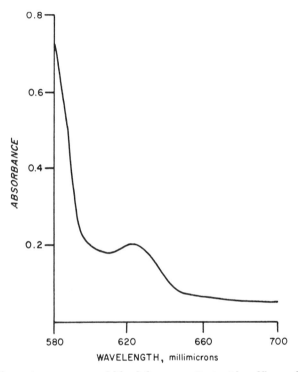

Fig. 5. Absorption spectrum of blood from a patient with sulfhemoglobinemia. Cyanide has been added to the diluted specimen.

When either methemoglobin or sulfhemoglobin is present in sufficient quantity, the brown pigment in blood is visible to the eye. Because these pigments do not oxygenate as does hemoglobin, the color of the arterial blood is not different from that of venous blood when their levels are sufficiently high to mask the color of oxyhemoglobin. Because the color of the skin is in part due to the blood flowing through the capillaries, the patient who has methemoglobinemia or sulfhemoglobinemia will appear darker, i.e., "cyanotic." This condition of color change is also present in the patient whose blood for other reasons is not properly oxygenated.[6]

Methemoglobinemia is usually caused by exposure to various chemical agents, many of which contain nitrogen. *The oxidation of the iron in hemoglobin from the ferrous to the ferric state produces methemoglobin.* In vitro, this process may be accomplished with oxidizing agents such as ferricyanide, or with nitrite, and may be reversed with sodium dithionite ($Na_2S_2O_4$). A substance has been reported as occurring in plasma, which promotes the spontaneous oxidation of hemoglobin to methemoglobin (20). In at least one instance this substance was identified as benzoquinoneacetic acid (21). Fatal methemoglobinemia has resulted from drinking water containing nitrites from wells that are contaminated with surface water exposed to animal waste material. Accidental use of nitrite instead of nitrate salts for food preservation has caused acute and fatal methemoglobinemia. Industrial exposure (either by inhalation or contact with skin) to nitro-organic compounds such as nitrobenzene and to aniline compounds readily produces methemoglobin.[7] Workers in the industries that manufacture such compounds should be regularly examined for evidence of methemoglobinemia.

Chronic use of certain drugs such as phenacetin, and related compounds, and certain sulfonamide derivatives may produce methemoglobinemia. Occasionally, sulfhemoglobin is produced in addition to methemoglobin after exposure to these agents. In some instances, sulfhemoglobin may be produced alone without significant increases in methemoglobin (Fig. 5). Whereas methemoglobin disappears within a few days after removal of the causative agent (6), sulfhemoglobin disappears much more slowly. This difference in the rate of disappearance can be explained on the basis of the presence of the methemoglobin reductase in normal red blood cells, which reconverts the methemoglobin to hemoglobin (23). Since sulfhemoglobin

[6] A gross test for abnormal pigments such as methemoglobin or sulfhemoglobin may be performed at the patient's bedside. Approximately 1 ml. of venous blood in a stoppered tube is aerated by inverting it for 1 minute; if no abnormal pigments are present the blood will become bright red. Failure of the blood to change to a bright red as compared with a normal control is an indication for quantitative studies.

[7] For additional information concerning causes of methemoglobinemia and sulfhemoglobinemia, see reference 22.

becomes a permanent part of the cell its disappearance is related to the red cell life span.

Hereditary methemoglobinemia is of special interest, although it is a rare condition. Gerald (24) described hereditary methemoglobinemia as a group of diseases which may be categorized by those in which there is: (1) an abnormal methemoglobin, (2) a deficiency in the red blood cell reducing system, and (3) an abnormal production of oxidizing substances in the blood. The methemoglobinemia of hemoglobin M is an example of the first category. Methemoglobin M differs from "normal" methemoglobin in its absorption spectrum and in its response to cyanide (25). Therefore, the presence of hemoglobin M invalidates the use of the spectrophotometric method for quantitation of methemoglobin. Under such conditions, incomplete reduction of the absorbance at 620 mμ upon the addition of cyanide indicates the presence of an abnormal pigment. A complete study of the continuous visible absorption spectra would be useful in identification of the pigment.

Normal Range[8]

Paul and Kemp (16) reported methemoglobin concentration of 0.6 to 2.5% of total hemoglobin in blood donors and hospital patients who were not receiving drugs that usually cause methemoglobin formation. This percentage range is in general agreement with ranges reported by others (14). The exact levels necessary to produce cyanosis or other clinical symptoms are not well established possibly because of variations between individual patients (22) as well as techniques of various analysts (6). Cyanosis may be observed in the range of 10 to 20% conversion of hemoglobin to methemoglobin and, in some patients, symptoms of anoxia may show above 30%; if the condition is chronic, tolerance to moderate concentrations of methemoglobin has been observed (6, 22, 26). The lethal level in man has not been determined but is probably in excess of 70% (26).[9]

[8] The spectrophotometric method is adequate to demonstrate abnormal levels of methemoglobin. With some variation of procedure based on the principles described, the precision and accuracy of this method should be adequate for use within the normal or low abnormal concentration ranges.

[9] The validity of identification of hemoglobin pigments in cadaver blood is open to question, particularly if sufficient time for bacterial decomposition to take place has elapsed since death. Under these conditions a number of pigments including methemoglobin may be formed (14, 25).

REFERENCES

1. Michel, H. O., and Harris, J. S., The blood pigments. The properties and quantitative determination with special reference to the spectrophotometric methods. *J. Lab. Clin. Med.* **25**, 445–463 (1940).
2. Van Slyke, D. D., and Hiller, A., Gasometric determination of methemoglobin. *J. Biol. Chem.* **84**, 205–210 (1929).
3. Heilmeyer, L., "Medizinische Spectrophotometrie." Fischer, Jena, 1933. (English translation by A. Jordan and T. L. Tippell, "Spectrophotometry in Medicine." Adam Hilger, London, 1943.)
4. Austin, J. H., and Drabkin, D. L., Spectrophotometric studies. II. Methemoglobin. *J. Biol. Chem.* **112**, 67–88 (1935).
5. Evelyn, K. A., and Malloy, H. T., Microdetermination of oxyhemoglobin, methemoglobin and sulfhemoglobin in a single sample of blood. *J. Biol. Chem.* **126**, 655–662 (1938).
6. Hamblin, D. O., and Mangelsdorff, A. F., Methemoglobin and its measurement. *J. Ind. Hyg. Toxicol.* **20**, 523–530 (1938).
7. Hutchinson, E. B., The measurement of methemoglobin. *Am. J. Med. Technol.* **26**, 75–83 (1960).
8. Deibler, G. E., Holmes, M. S., Campbell, P. L., and Gans, J., Use of Triton X-100 as a hemolytic agent in the spectrophotometric measurement of blood O_2 saturation. *J. Appl. Physiol.* **14**, 133–136 (1959).
9. Gambino, S. R. Personal communication (1963).
10. Hainline, A., Jr., Hemoglobin. *In* "Standard Methods of Clinical Chemistry" (D. Seligson, ed.), Vol. 2, pp. 49–60. Academic Press, New York, 1958.
11. Cannon, R. K., Proposal for a certified standard for use in hemoglobinometry. Second and final report. *Clin. Chem.* **4**, 246–251 (1958).
12. Connerty, H. V., and Briggs, A. R., New method for the determination of whole blood and hemoglobin. *Clin. Chem.* **8**, 151–157 (1962).
13. Zijlstra, W. G., and VanKampen, E. J., Standardization of hemoglobinometry. I. The extinction coefficient of hemoglobincyanide. *Clin. Chim. Acta* **5**, 719–726 (1960).
14. Lemberg, R., and Legge, J. W., "Hematin Compounds and Bile Pigments." Wiley (Interscience), New York, 1949.
15. Meites, S., and Faulkner, W. R., "Manual of Practical Micro and General Procedures in Clinical Chemistry." Thomas, Springfield, Illinois, 1962.
16. Paul, W. D., and Kemp, C. R., Methemoglobinemia, a normal constituent of blood. *Proc. Soc. Exptl. Biol. Med.* **56**, 55–56 (1944).
17. Schwerd, W., "Der rote Blutfarbstoff und seine wichtigsten Derivate," Max Schmidt-Römhild, Lübeck, 1962.
18. Drabkin, D. L., and Austin, J. H., Spectrophotometric studies: Preparations from washed blood cells: nitric oxide hemoglobin and sulfhemoglobin. *J. Biol. Chem.* **112**, 51–65 (1935).
19. Horecker, B. L., and Brackett, F. S., A rapid spectrophotometric method for the determination of methemoglobin and carboxyhemoglobin in blood. *J. Biol. Chem.* **152**, 669–677 (1944).
20. Homolka, J., Increase in oxidizing agents in serum causing methaemo-

globinaemia and a test for this condition. *Clin. Chim. Acta* **3**, 603–604 (1958).

21. Fishberg, E. H., Excretion of benzoquinone acetic acid in hypovitaminosis C. *J. Biol. Chem.* **172**, 155–163 (1948).

22. Finch, C. A., Methemoglobinemia and sulfhemoglobinemia. *New Engl. J. Med.* **239**, 470–478 (1948).

23. Huenekens, F. M., Caffrey, R. W., Basford, R. E., and Gabrio, B. W., Erythrocyte metabolism, IV. Isolation and properties of methemoglobin reductase. *J. Biol. Chem.* **227**, 261–272 (1957).

24. Gerald, P. S., The hereditary methemoglobinemias. *In* "The Metabolic Basis of Inherited Disease" (J. B. Stanbury, J. B. Wyngaarden, and D. S. Frederickson, eds.), Chapter 33, pp. 1068–1085. McGraw-Hill, New York, 1960.

25. Gerald, P. S., The electrophoretic and spectroscopic characterization of hemoglobin M. *Blood* **13**, 936–949 (1958).

26. Dubowski, K. M., Measurements of hemoglobin derivatives. *In* "Hemoglobin. Its Precursors and Metabolites" (F. W. Sunderman and F. W. Sunderman, Jr., eds.), Chapter 5, pp. 49–60. Lippincott, Philadephia, Pennsylvania, 1964.

OSMOLALITY OF SERUM AND URINE*

Submitted by: ROY B. JOHNSON, JR., Scripps Clinic and Research Foundation, La Jolla, California, and HANS HOCH, Veterans Administration Center, Martinsburg, West Virgina

Checked by: WERNER R. FLEISCHER, St. Joseph Hospital, Joliet, Illinois
FRANK A. IBBOTT, University of Colorado Medical Center, Denver, Colorado
MIRIAM REINER, District of Columbia General Hospital, Washington, District of Columbia

Introduction

The importance of osmotic pressure for living cells is well known; it is one of the main factors regulating the homeostatic equilibrium between cytoplasm and extracellular fluid. Osmotic pressure is often considered merely a property of concentration and therefore provides information interchangeable with other properties like specific gravity and refractive index. Confusion exists because this is partly true.

NOTE: *Specific gravity* depends on the total weight of dissolved material in a unit volume. On a molar basis, heavier elements contribute more than lighter elements. *Refractive index* of a solution depends on the concentration of dissolved material, but the change in the refractive index produced differs according to the optical properties of the solutes.

In contrast, *osmotic pressure* is uniquely a measure of the summation effect of all the particles in solution and does not discriminate between molecules, ions, aggregates, heavy or light component elements, and those with differing optical properties. The reason for this is that *osmotic pressure reflects the deficit from 100%* (or mole fraction 1) *in the concentration of water* rather than reflecting directly the concentration of substances dissolved. A clear relationship does exist between specific gravity, refractive index, and osmotic pressure when the solution composition is known. But, because biological fluids are complex mixtures of often inexactly known composition, this relationship can only be used to approximate the osmotic pressure.

Osmotic pressure should not be confused with *colloid osmotic pressure* (oncotic pressure) which is a measure of the number of particles (proteins, for example) that do not pass through cell membranes. These membranes are semipermeable in different degrees, but never retentive for all dissolved substances (especially crystalloids), so that the actual excess pressure in the cell is never equal to the osmotic

* Based on modification of the work of Beckmann (1).

159

pressure. Nevertheless, osmotic pressure determinations serve a useful purpose with a potential which is yet to be fully appreciated.

The term "osmolality" has been conceived for biological applications and defined as the molality of an ideal substance dissolved in water in amounts sufficient to produce the same osmotic pressure and the same freezing point depression as the specimen produces (2). This ideal substance is assumed to be a nonelectrolyte, neither dissociating nor associating, and not interacting with the water. The amount of dissolved material equivalent in effect to 1 mole of the ideal substance is called 1 Osmol.

Principle

The osmolality of serum or urine is determined by comparing its freezing point with that of a NaCl solution of known osmotic pressure.

Theory

The formation of a homogeneous mixture by the addition of any material (solute) to a pure liquid (solvent) produces changes in several characteristics of the solvent. Four of these changes, sometimes referred to as the colligative properties, can be shown to be mathematically interrelated, and in dilute solutions these changes are linear as the solute concentration increases:

1. The freezing point is lowered.
2. The boiling point is raised.
3. The vapor pressure is lowered.
4. The osmotic pressure is increased.

The Submitters will not discuss in detail the thermodynamics of these relationships. Interested persons should refer to treatises on physical chemistry (3). The lowering of vapor pressure is the hinge upon which the others swing, and in dilute solutions this is proportional to the mole-fraction of the solute in the mixture (Raoult's law). Mole fractions can be calculated more easily from molalities (gram-mole/kilogram solvent) than from molarities (gram-mole/liter solution). Although, for aqueous solutions, the two approach each other at 4°C. as the solute concentration approaches zero, the term "osmolarity" should be discarded in favor of the term "osmolality" because osmolality bears a closer relationship to mole-fraction.

Electrolytes can contribute two or more times as many particles (ions) for each gram-molecular weight as do nonionizing substances. Except for very dilute solutions, electrolytes, even if completely ionized, do not exert the full effects theoretically expected from the number of particles because of electrostatic attraction between the ions. The ratio of the actual colligative property (freezing point lowering, for example) to the theoretical one expected from the same

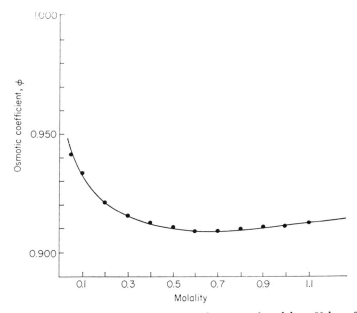

FIG. 1. Osmotic coefficient of NaCl as a function of molality. Values for ϕ equal $1 - j$ were obtained from j values given in reference (4).

number of noninteracting particles, is called the osmotic coefficient (ϕ). As the latter is a function of the concentration, electrolyte solutions of known freezing point cannot be diluted with solvent to produce a new solution with proportionally smaller freezing point depression; although, in the case of a known solute, the freezing point depression can be predicted from the calculated molality of the diluted solution. The osmotic coefficient of NaCl plotted against molality for the range of concentrations pertinent to clinical work is presented in Fig. 1.

The freezing point depression (ΔT) produced by an ideal nonionic solute is given for dilute solutions by:

$$\Delta T = \frac{RT^2}{H_{\text{fus}}} \cdot N_2, \tag{1}$$

where R is the gas constant equal to 1.987 cal. deg.$^{-1}$ mole^{-1}, T is the absolute freezing temperature of the solvent, N_2 is the mole-fraction of solute, and H_{fus} is the molar heat of fusion of the solvent. For water the molal depression constant, $\Delta T/m$, is 1.858°C. per unit molality.

In biological systems it is conventional to express the osmolality in milliosmoles per kilogram water. Thus 1 mOsmol. kg.$^{-1}$ gives a freezing point depression of 0.001858°C.

Method

The osmolality of a biological fluid can be determined by measuring either freezing point, vapor pressure, or osmotic pressure. The determination of freezing point has been chosen for clinical work because of its simplicity. Devices are also available for measuring vapor pressure, but are still in need of proper evaluation (5).

The freezing point of a solution may be defined as that temperature at which an infinitesimal amount of solid phase exists in equilibrium with the liquid phase, at standard pressure (6). In practice, the water in biological fluids will not start crystallizing at the freezing point, and it is necessary to supercool as well as to initiate crystallization either by "seeding" or by some other means, such as a vibrating rod. During crystallization, which is almost instantaneous once started, heat is liberated and the temperature rises until the freezing point of the now slightly more concentrated solution is reached (about 3% of the total water freezes for 2° of supercooling). The different methods of correcting for the error in the freezing point caused by supercooling, and the rationale for the technique used in modern osmometry, are described in reference 6.

Procedure

1. *Standards*. A widely used but slightly incorrect set of NaCl solutions for calibration purposes is given by Fiske (7). The amounts of NaCl required for preparing this set of solutions of defined osmolality are listed in Table I for historical and comparative purposes

TABLE I

EXPECTED FREEZING POINT LOWERING AND COMPOSITION OF NaCl STANDARDS
ACCORDING TO FISKE (7)

Stated value of standard (milliosmoles/ kg.-H$_2$O)	Revised value of standard[a] (milliosmoles/ kg.-H$_2$O)	Expected freezing point lowering (°C.)	Sodium chloride (g./kg. H$_2$O)
100[b]	99.84	0.186	3.089
300[b]	299.4	0.557	9.457
400	400.0	0.743	12.70
500[b]	500.0	0.929	15.93
750	752.2	1.394	24.10
1000[b]	1003	1.86	32.23
1200	1206	2.23	38.76
1400	1407	2.60	45.22
1600	1608	2.97	51.62
1800	1809	3.34	58.01
2000	2012	3.72	64.37
2500	2517	4.65	79.97
3000	3024	5.57	95.40

[a] Calculated from (4).
[b] Most needed.

only. (Corrected values of these standards are given in column 2.) They were calculated from Eq. (2):

$$C = 0.1086 \frac{O}{K} \qquad (2)$$

where C is the number of grams of NaCl to be dissolved in 1 kg. water, O is the required value for the milliosmolality, and K is the freezing point lowering per unit molality for NaCl at the particular milliosmolality. K is given in degrees C./moles per kilogram H$_2$O. The numerical constant, 0.1086, is one-thousandth of the product of the molecular weight of NaCl and the molal freezing point depression constant of water, 1.858°C.

Equation 2 above is the rearranged statement for the definition of osmolality:

$$\frac{O}{1000} = \frac{\overbrace{(1.858n\phi)}^{K}(C/58.45)}{1.858} \qquad (3)$$

where $O/1000$ is the osmolality and the numerator on the right is the actual freezing point depression obtained at a molality of $C/58.45$ moles NaCl per kilogram H_2O; n is the theoretical number of particles contributed by each molecule, equal to 2 for NaCl; and ϕ is the osmotic coefficient. The values for K in the Fiske Table are taken from the International Critical Tables (8) and from "The Handbook of Chemistry and Physics" (9), now outdated.

The Submitters recalculated the amounts of NaCl required, using more recent data (4). These are listed in Table II. There is close

TABLE II

EXPECTED FREEZING POINT LOWERING AND COMPOSITION OF NaCl STANDARDS
CALCULATED FROM REFERENCE (4)

Standard	Expected freezing point lowering (°C.)	Recalculated for nominal value of Fiske (g. NaCl/kg.)
100	0.186	3.094
300	0.557	9.476
400	0.743	12.70
500	0.929	15.93
750	1.394	24.03
1000	1.86	32.12
1200	2.23	38.57
1400	2.60	44.98
1600	2.98	51.37
1800	3.35	57.72
2000	3.72	63.97 ± 0.3[a]
2500	4.65	79.44 ± 0.8[a]
3000	5.58	94.64 ± 1.5[a]

[a] Range of uncertainty of interpolation.

agreement between the two sets of standards except at high concentrations. The data in Table II are more correct and the Submitters recommend their adoption at this time.

Dissolved atmospheric gases will alter the freezing points of standards and specimens. The changes are small and they are ignored in clinical work.

As NaCl is not hygroscopic, and reagent grades contain very little water (H. H. found one batch with less than 0.1%), it is convenient to use the reagents as supplied and to determine the water content on an aliquot of 1–2 g. from a newly opened bottle by heating overnight to anywhere between 200° and 500°C., or to the beginning of melting for not longer than 1 minute in a covered platinum crucible. Weigh the correct amount of NaCl and add to 1 kg. of water (double glass-distilled) contained in a volumetric flask bearing a calibration mark for 1 kg. at a specified temperature.

NOTE: To calibrate a volumetric flask weigh a clean dry 1 l. volumetric flask on a balance with a sensitivity better than 10 mg. and capable of carrying 2 kg. Fill the flask with double glass-distilled water until it is balanced against 1 kg. more than the weight of the flask. Use a diamond point to mark the meniscus and record the exact temperature of the water.

Alternatively, fill a Class A, 1 l. volumetric flask to the mark at 20°C. and add 1.8 ml. water before adding the salt (W.R.F.).

Store the standard solutions in clean polyethylene or borosilicate bottles. Cap the stoppers and exposed lips of the bottles with Parafilm to prevent contamination by dust. As some polyethylene bottles may not close tightly, test them by inverting and applying pressure. Obtain samples by first *pouring* sufficient amounts into clean tubes and never by pipetting directly from the bottle.

2. *Apparatus.* Instruments employing thermistor probes in a null meter circuit are available from several manufacturers.[1] Incorporation of a pulsator to initiate freezing permits repeated determinations which cannot be done after seeding. The reader should follow the directions provided by the manufacturer for operating details of his particular instrument.

3. *Obtaining and storing specimens.* Collect the blood by venipuncture with minimal stasis. Use only clean and dry needles, syringes, and tubes. Rim the clot, centrifuge, and separate the serum soon after collection. Hemolysis should be avoided, though it does not usually interfere because lysed red cells and serum have nearly the same osmolality (10). The Submitters recommend a second centrifugation to obtain serum free of particulate matter which could act as a seeding agent and cause crystallization before the correct degree of supercooling is attained. Cap securely and refrigerate or freeze serum that cannot be analyzed immediately. Avoid delay in the determination because storage lowers the result (10). Specimens re-

[1] Advanced Instruments Inc., 45 Kenneth St., Newton Highlands 61, Massachusetts; American Instrument Co., Inc., 8030 Georgia Ave., Silver Springs, Maryland; Fiske Associates, Inc., 186 Greenwood Ave., Bethel, Connecticut; Precision Instruments, 6 Cornell Road, Framingham, Massachusetts.

main unchanged for 3 hours at room temperature and for 10 hours at 4°C. (M.R.).

Collect urine in sterile, clean, dry glassware without preservatives. Cap the bottle to avoid evaporation. Regulation of the patient in various ways prior to and during the collection period has been described (11–15). These precautions permit assessment of different aspects of renal physiology. Make determinations on freshly voided and centrifuged urine to avoid errors from changes in composition caused by chemical or bacterial action. Centrifuge the specimen enough to give a crystal clear supernate. Redissolve cold-precipitated substances by warming refrigerated specimens before centrifuging.

4. *Calculations.* Most instruments are provided with a direct read-out attached to the potentiometer in the null-point circuit. Error in the read-out, as determined experimentally with the standard solutions, can often be corrected by adjustment of the instrument according to the manufacturer's directions. Otherwise, graphical or arithmetical methods can be used to obtain corrections from readings on the standard solutions.

Discussion

With modern instrumentation the analyst can obtain the osmolal concentration of body fluids in a few minutes. Specimens are usually unchanged after the determination and, therefore, can be used for other analyses. Although 2 ml. of specimen is the customary amount required, most instruments can be adapted to measure the freezing point of 0.2 ml.

It is possible to obtain reproducibility to better than ±1 mOsmol./kg. and an equal accuracy if the standards are correct. Failure can usually be traced to particulate matter in the specimen, inattention to the standardization of the time-cooling sequence, improper care and filling of the cooling bath (if present), or adulteration by dirt or moisture on glassware and probe.

Available data on normal values of serum and urine are controversial. The normal range for serum osmolality has been reported to be 289 ± 8 mOsmol./kg. (11) and 289–308 mOsmol./kg. (12).

NOTE: The Submitters favor the values about the mean of 289. For comparison, normal saline (0.9%, w/v) has an osmolality of 285 mOsmol./kg.

A formula (16) to correct high values downward in hyperglycemic and azotemic states should not be used. This practice repudiates the basic purpose of the test, which is to measure osmotic pressure and not a particular solute as, for example, sodium ion.

The normal range of urine osmolalities is broad. The osmotic limits of renal dilution and concentration have been reported to be 40 and 1400 mOsmol./kg., respectively (17). During maximal urine concentration the normal range of urine osmolality has also been reported to be 967 to 1342 (12), and 855 to 1335 mOsmol./kg. (18).

NOTE: The Submitters favor this last range.

Urine osmolality determinations will yield values outside this range in cases of advanced renal disease or extrarenal disease affecting urine concentration (W.R.F.).

The osmolality of spinal, pleural, and peritoneal fluids is equal to or little different from that of serum drawn at the same time (M.R.; 10).

Serum osmolality studies have been used in the evaluation of hyper- and hyponatremia, renal solute retention in acute renal failure, and hydration status (W.R.F.). Serum studies are also a means of detecting "undetermined solute" in cases of poisoning and of estimating the requirement for and effectiveness of dialysis (W.R.F.). Most importantly, they are used to estimate serum water since the concentration of the water determines the osmotic pressure.

Urine osmolality and urine/serum osmolality ratios have found application in the calculation of free water and osmolar clearances in the differential diagnosis of polyuria and in the evaluation of renal solute excretion in certain hormonal and renal diseases (W.R.F.).

REFERENCES

1. Beckmann, E., Über die Methodik der Molekulargewichtsbestimmung durch Gefrierpunktserniedrigung. Z. physik. Chem. 3, 638 (1888).
2. Wolf, A. V., "Thirst," p. 465. Thomas, Springfield, Illinois, 1958.
3. Daniels, F., and Alberty, R. A., "Physical Chemistry," pp. 219–230. Wiley, New York, 1955; Maron, S. H., and Prutton, C. F., "Principles of Physical Chemistry," pp. 183–207. Macmillan, New York, 1959.
4. Landolt, H. u. Bornstein, R., "Zahlenwerte und Funktionen aus Physik, Chemie, Astronomie, Geophysik und Technik" (J. Bartels, P. Ten Bruggencate, H. Hausen, K. H. Hellwege, Kl. Schäfer, and E. Schmidt, eds.), 6th ed., Vol. 2a, p. 865. Springer, Berlin, 1960; Scatchard, G., and Prentiss, S. S., The freezing points of aqueous solutions. IV. Potassium, sodium and lithium chlorides and bromides. J. Am. Chem. Soc. 55, 4358–4359 (1933).
5. Mechrolab, Inc., 1062 Linda Vista, Mountain View, California; Rosemont Engineering Co., 4900 W. 78th St., Minneapolis 24, Minnesota.

6. Abele, J. E., The physical background to freezing point osmometry and its medical-biological applications. *Am. J. Med. Electronics* **2**, 32–41 (1963).
7. Osmometer Newsletter No. 4, Fiske Assoc., Inc., Bethel, Connecticut, 1962.
8. "International Critical Tables," Vol. 4, p. 258. McGraw-Hill, New York, 1928.
9. Hodgman, C. D., Weast, R. C., and Selby S. M., (eds.), "The Handbook of Chemistry and Physics," 40th ed., p. 2299. Chemical Rubber Co., Cleveland, Ohio, 1958.
10. Hendry, E. B., The osmotic pressure and chemical composition of human body fluids. *Clin. Chem.* **8**, 246–265 (1962).
11. Hendry, E. B., Osmolarity of human serum and of chemical solutions of biologic importance. *Clin. Chem.* **7**, 156–164 (1961).
12. Lindemann, R. D., Van Buren, H. C., and Raisz, L. G., Osmolar renal concentrating ability in healthy young men and hospitalized patients without renal disease. *New Engl. J. Med.* **262**, 1306–1309 (1960).
13. Baldwin, D. S., Berman, H. J., Heinemann, H. O., and Smith, H. W., The elaboration of osmotically concentrated urine in renal disease. *J. Clin. Invest.* **34**, 800–807 (1955).
14. Dreifus, L. S., Frank, M. N., Bellet, S., Determination of osmotic pressure in diabetes insipidus. *New Engl. J. Med.* **251**, 1091–1094 (1954).
15. Zak, G. A., Brun, C., and Smith, H. W., The mechanism of formation of osmotically concentrated urine during the antidiuretic state. *J. Clin. Invest.* **33**, 1064–1074 (1954).
16. Edelman, I. S., Leibman, J., O'Meara, M. P., and Birkenfeld, L. W., Interrelations between serum sodium concentration, serum osmolality and total exchangeable sodium, total exchangeable potassium and total body water. *J. Clin. Invest.* **37**, 1236–1256 (1958).
17. Wolf, A. V., "The Urinary Function of the Kidney." Grune & Stratton, New York, 1950.
18. Jacobson, M. H., Levy, S. E., Kaufman, R. M., Gallinek, W. E., and Donnelly, O. W., Urine osmolality. *Arch. Internal Med.* **110**, 83–89 (1962).

Pertinent Articles and Books

19. Albrink, M. J., Hald, P. M., Man, E. B., and Peters, J. P., The displacement of serum water by the lipids of hyperlipemic serum. A new method for the determination of serum water. *J. Clin. Invest.* **34**, 1483–1488 (1955).
20. Christensen, H. N., "Body Fluids and Their Neutrality." Oxford Univ. Press, London and New York, 1963.
21. Gamble, J. L., "Chemical Anatomy Physiology and Pathology of Extracellular Fluid." Harvard Univ. Press, Cambridge, Massachusetts, 1952.
22. Singer, D. L., Drolette, M. E., Hurwitz, D., and Freinkel, N., Serum osmolality and glucose in maturity onset diabetes mellitus. *Arch. Internal Med.* **110**, 758–762 (1962).
23. Moyer, J. H., and Fuchs, M. (eds.), "Edema, Mechanism and Management," Hahneman Symposium on Salt and Water Retention, Saunders, Philadelphia, Pennsylvania, 1960.

See also "Bibliography of Articles Describing Experimental Work in Osmometry." Fiske Assoc., Inc., Bethel, Connecticut.

pH[*] and P_{CO_2}

Submitted by: S. RAYMOND GAMBINO, The Englewood Hospital, Englewood,
New Jersey; and Columbia-Presbyterian Medical Center, New
York, New York
Checked by: GEORGE D. BATES, The Children's Hospital, Columbus, Ohio
WERNER R. FLEISCHER, St. Joseph Hospital, Joliet, Illinois
T. C. HUANG, Timken Memorial Hospital, Canton, Ohio
RALPH E. THIERS, Duke University Medical Center, Durham,
North Carolina

Introduction

The hydrogen ion activity of mammalian blood was first measured in
1903 by Höber (1), who defibrinated a fresh sample of ox blood

[*] Readers should refer to the chapter on "Determination of Blood pH" in
Volume 2 of this series (51) as a prelude to this article. (*Ed.*)

169

equilibrated with 4.3% CO_2. The hydrogen ion activity measurement was made with a hydrogen electrode and the result was equivalent to a pH of 7.23. A precise measurement of mammalian blood pH was made by Hasselbalch and Lundsgaard (2) in 1912. They analyzed four samples of ox blood at 38.5°C. and a P_{CO_2} of 40 mm. Hg, and obtained a mean pH of 7.36 with a range of 7.31 to 7.45.

In 1917, Parsons (3) made a careful electrometric pH measurement of blood by using a precise hydrogen electrode and an electrostatic potentiometer. He obtained *10 ml.* of capillary blood from his finger by making a deep puncture with a lancet and he defibrinated this blood with a feather. The blood was kept on ice in a sealed container until needed, and Parsons separated serum from cells at body temperature using a specially designed anaerobic thermos-bottle flask. His capillary blood pH was 7.38 at a P_{CO_2} of 40 mm. Hg, and he calculated his venous pH to be 7.36.

The glass electrode was first used for human blood pH measurement in 1925 by Kerridge (4), who obtained a pH of 7.42. The potential developed across the glass membrane was read with an electrostatic potentiometer, but moisture and dirt caused serious problems because of the high impedance of glass.

Stadie (5) designed an electron tube potentiometer in 1929 for determination of blood pH with a glass electrode. He later (6) measured serum pH at 38°C. with an improved thermostated glass electrode and an improved potentiometer, and obtained pH values ranging between 7.36 and 7.47 at a P_{CO_2} of 40 mm. Hg. In addition, he compared the glass electrode with the hydrogen electrode and found the average glass electrode reading to be 0.007 pH units higher.

The Sanz (7) and Astrup (8) micro capillary measuring chains are miniaturized versions of the classic flow-through electrode developed by MacInness and Belcher (9). The pH meters used with micro capillary chains are more sophisticated versions of the classic Stadie electron tube potentiometer.

NOTE: Colorimetric methods of measuring blood pH were once widely used (10), but these methods have since been replaced by the simpler glass electrode techniques. The colorimetric methods, however, are not inaccurate. Van Slyke (11) measured the pH of 26 plasma samples with a colorimetric and a hydrogen electrode technique. Twenty one of the 26 samples differed by less than 0.02 pH units. The greatest difference was 0.04 pH units, and the average difference between the colorimetric and hydrogen electrode results was only 0.002 pH units.

P_{CO_2}, the partial pressure of carbon dioxide, can be determined in several ways:

(a) Calculation from bood or plasma pH and plasma total CO_2 content (12).

(b) Interpolation from changes in pH after equilibration with gases of known P_{CO_2} (8).

(c) Direct measurement with a P_{CO_2} electrode (13).

(d) Measurement of end tidal alveolar CO_2 concentration (14, 15).

(e) Equilibration with an air-bubble in the Scholander apparatus (16).

(f) Rebreathing expired air with analysis of final CO_2 concentration at the end of the rebreathing period (17).

Only the first three methods are discussed in this chapter.

Principles

pH

If a thin glass membrane is inserted between two solutions of unequal hydrogen ion concentration, a potential will develop across the glass (18). In the standard glass electrode the solution on one side of the glass membrane is kept sealed and constant. This solution is usually 0.1 N HCl. Electrical contact with the 0.1 N HCl solution is made with a silver-silver chloride electrode which in turn is connected to the input of the pH meter through a shielded copper wire. The solution whose pH is to be measured is placed on the unsealed side of the glass membrane. Electrical contact with the solution to be measured is made through saturated KCl and a calomel reference electrode. The lead from the reference electrode is returned to the pH meter to complete the measuring circuit. The complete measuring chain is shown diagrammatically in Fig. 1.

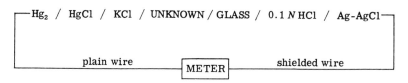

FIG. 1. Components of a complete pH measuring chain.

A potential is developed at every junction between two different solids or solutions in the measuring chain. The only potential, however, that can be permitted to vary from measurement to measurement is that arising between the unknown and the glass. Proper circuit design, temperature control, grounding, and shielding all insure constancy of the other potentials.

Accurate temperature control is extremely important. A change in electrode temperature will have two independent effects on pH measurements. A change in electrode temperature will vary the amount of voltage generated across the glass membrane by a particular change in hydrogen ion concentration. For example, at 25°C. a change in hydrogen ion concentration of 1 pH unit will produce a voltage change of 59.15 millivolts; while at 100°C. the same 1 pH unit change will produce a voltage change of 74.03 millivolts. This variation in voltage change with change in temperature is constant and is compensated for by the *temperature dial* found on all pH meters.

The second, and independent, change occurring with a change in electrode temperature is an actual alteration of hydrogen ion activity in the solution being measured. The variation of solution pH with temperature is a unique property of each solution and may vary in a positive or negative direction, or not at all. This temperature effect on pH is not compensated for by the temperature dial on the pH meter.

CALCULATED P_{CO_2}

P_{CO_2} can be estimated by solving the Henderson-Hasselbalch equation for one unknown (19).

$$pH = pK + \log \frac{HCO_3^- \text{ (bicarbonate)}}{H_2CO_3 \text{ (carbonic acid)}}$$

$$pK = 6.10 \qquad\qquad (20, 21)$$

Carbonic acid and P_{CO_2} are interchangeable on the basis of the solubility factor, "S."

$$P_{CO_2} \times S = H_2CO_3$$
$$S = 0.0308 \text{ at } 37.0°C.$$
$$0.0304 \text{ at } 37.5°C.$$
$$0.0301 \text{ at } 38.0°C. \qquad (19)$$

The total CO_2 content of plasma represents the sum of bicarbonate plus carbonic acid. Bicarbonate can be derived from total CO_2 content by subtracting carbonic acid, or $S \times P_{CO_2}$. The Henderson-Hasselbalch equation can then be rewritten as:

$$pH = 6.10 + \log \left[\frac{(CO_2 \text{ content}) - (0.03 \times P_{CO_2})}{0.03 \times P_{CO_2}} \right]$$

and by further arrangement:

$$P_{CO_2} = \frac{CO_2 \text{ content}}{0.03 \times [\text{antilog}(pH - 6.10) + 1]}$$

Nomograms and tables based on the above formula are available for estimation of P_{CO_2} from pH and total CO_2 content (22, 23, 24, 25). The Weisberg table (25) is simple to use and is reproduced in Table I. This table is calculated for a temperature of 37°C. and a pK of 6.10. A change in body temperature will alter S and pK, and a change in pH will alter pK (19, 20).

TABLE I

REVISED TABLE OF FACTORS (f_P) OBTAINED FROM pH TO CALCULATE PLASMA (OR SERUM) PARTIAL PRESSURE OF CO_2 (P_{CO_2}) FROM THE TOTAL CO_2 CONTENT

$$P_{CO_2} = f_P \times [CO_2 \text{ content}]$$
$$[H_2CO_3] = P_{CO_2} \times 0.03$$
$$[HCO_3^-] = [CO_2 \text{ content}] - [H_2CO_3]$$

P_{CO_2} Factors (f_P)

pH	0.00	0.01	0.02	0.03	0.04	0.05	0.06	0.07	0.08	0.09
6.8	6.42	6.30	6.15	6.00	5.85	5.72	5.60	5.48	5.36	5.22
6.9	5.10	5.00	4.88	4.76	4.66	4.54	4.44	4.32	4.24	4.16
7.0	4.05	3.95	3.88	3.78	3.70	3.62	3.52	3.44	3.38	3.30
7.1	3.21	3.15	3.08	3.00	2.92	2.86	2.80	2.74	2.68	2.61
7.2	2.55	2.50	2.44	2.38	2.33	2.27	2.22	2.16	2.12	2.08
7.3	2.03	1.98	1.94	1.89	1.85	1.81	1.76	1.72	1.69	1.65
7.4	1.61	1.58	1.54	1.51	1.46	1.43	1.40	1.37	1.34	1.31
7.5	1.28	1.25	1.22	1.19	1.16	1.14	1.11	1.08	1.06	1.04
7.6	1.02	1.00	0.97	0.95	0.92	0.90	0.88	0.86	0.84	0.83
7.7	0.81	0.79	0.77	0.75	0.73	0.72	0.70	0.69	0.67	0.66
7.8	0.64									

INTERPOLATED P_{CO_2}

The relationship of pH to log P_{CO_2} is linear for any particular blood sample (2). Therefore, if a pH/log P_{CO_2} plot can be obtained, and the actual pH is known, P_{CO_2} can be estimated. In practice, the actual pH of whole blood is measured first. Then two aliquots are equilibrated with two different concentrations of CO_2 gas. The concentrations of these gases must be known with great accuracy. The pH of each aliquot is measured after suitable equilibration and a plot of measured pH versus log P_{CO_2} is made for each aliquot. Since the relationship between pH and log P_{CO_2} is linear, a straight line is drawn between the two derived points. The actual P_{CO_2} of the original blood sample is obtained by locating the actual pH of the blood on the line (26).

DIRECT MEASUREMENT OF P_{CO_2}

P_{CO_2} can be measured with great accuracy and precision with a P_{CO_2} electrode (13, 27). A P_{CO_2} electrode is a modified pH electrode. Blood, plasma, or gas is introduced into a temperature controlled chamber, one wall of which is formed by a thin (0.001 inch) Teflon membrane. Small uncharged molecules (e.g., CO_2 gas) will readily traverse this membrane, but charged particles (e.g., hydrogen ions) will not. A thin layer (0.002 inch) of 0.1 M bicarbonate is placed on the opposite side of the membrane. The pH of this bicarbonate solution will vary with changes in carbonic acid concentration and thus with P_{CO_2}. The P_{CO_2} of the bicarbonate solution will vary directly with the P_{CO_2} of the sample because the Teflon membrane is completely and rapidly permeable to CO_2 gas. The pH of the bicarbonate solution is measured with a sensitive and stable glass electrode system. The pH change is converted into a P_{CO_2} reading through the use of a linear calibration curve relating log P_{CO_2} to pH. Some instruments have a logarithmic scale for direct reading of P_{CO_2}. The actual pH of the blood sample will have no effect on the pH of the bicarbonate solution because the membrane is impermeable to hydrogen ions.

Reagents

1. *Buffers.* The primary standard for blood pH measurements has been defined by the National Bureau of Standards (28). Make the primary standard buffer (pH 7.384 at 38°C) as follows:

Dissolve 1.179 g. (air weight) of potassium dihydrogen phosphate, KH_2PO_4, and 4.302 g. (air weight) of disodium hydrogen phosphate, Na_2HPO_4, in ammonia and carbon dioxide free distilled water and dilute to 1000 ml. at 25°C.[1]

The assigned pH values of this primary standard at different temperatures are:

7.429 at 20°C.	7.389 at 35°C.
7.413 at 25°C.	7.384 at 38°C.
7.400 at 30°C.	7.380 at 40°C.

[1] The phosphate salts are obtainable as certified standard samples from the National Bureau of Standards. The catalog numbers are: 186 Ib (KH_2PO_4) and 186 IIb (Na_2HPO_4). Information may be obtained by writing to the Standard Samples Unit, National Bureau of Standards, Washington 25, D.C.

The ionic strength of this primary buffer is 0.1, the potassium dihydrogen phosphate is 0.008695 molal, and the disodium hydrogen phosphate is 0.03043 molal.

A second buffer with a different pH (pH 6.840 at 38°C) is required to check the pH/millivolt response of the electrode system. Make the recommended National Bureau of Standards buffer as follows:

Dissolve 3.40 g. of potassium dihydrogen phosphate, KH_2PO_4, and 3.53 g. of disodium hydrogen phosphate, Na_2HPO_4, in ammonia and carbon dioxide free distilled water and dilute to 1000 ml. at 25°C. The assigned pH values of this second buffer are:

6.88 at 20°C.	6.84 at 35°C.
6.86 at 25°C.	6.84 at 40°C.
6.85 at 30°C.	

NOTE: The stability of phosphate buffers is variable. The National Bureau of Standards specifications make no mention of stability nor do they suggest any preservative. Radiometer[2] certifies its phosphate buffers in unopened bottles for only 6 months. Bottles are not certified once they are opened. Checker (W. R. F.) adds several granules of thymol to the buffers and refrigerates them. The phosphate buffers are then stable for at least 2 months. The primary cause of deterioration is the growth of organisms, particularly fungi.

NOTE: The Submitter uses a p-nitrophenol (PNP) buffer (pH 6.985 at 38°C.) and an acetate buffer (pH 4.64 at 38°C.).[3] Both buffers are stable without the addition of preservatives and without refrigeration. The Submitter has found the acetate and PNP buffers stable at least 6 months after opening a bottle, and stable at least 2 years in unopened bottles stored at 20° to 30°C. The PNP buffer is yellow and the acetate buffer is colorless, making differentiation simple and cross contamination easily detected. The PNP buffer has a temperature coefficient similar to that of plasma, while the acetate buffer has a negligible temperature coefficient. This difference in temperature coefficients makes it possible to detect thermostating errors. Finally, the wide pH separation of these

[2] Radiometer, 72 Emdrupvej, Copenhagen NV, Denmark, manufactures two precision buffers with nominal values of 6.84 and 7.38. These buffers are standardized according to the method of Spinner and Petersen (29) and thus are in close agreement with the National Bureau of Standards buffers. The pH of a buffer solution containing 1.816 g. of potassium dihydrogen phosphate and 9.501 g. of disodium hydrogen phosphate·$2H_2O$, per 1000 g. of water was found to be 7.381 ± 0.002 at 38.0°C. (29). These buffers are available from The London Co., 811 Sharon Dr., Westlake, Ohio.

[3] Metrohm AG, Herisau, Switzerland, manufactures two buffers with nominal values of 4.64 and 6.99 at 37°C. The 6.99 buffer is composed of equal volumes of 0.04 M p-nitrophenol and 0.04 M sodium p-nitrophenolate. The 4.64 buffer is composed of 200 ml. of 1 N acetic acid and 100 ml. of 1 N sodium hydroxide added to 1000 ml. of distilled water (30). These buffers are available from Brinkmann Instruments, Cantiague Rd., Westbury, New York.

buffers gives a sensitive and reliable check of the pH/millivolt response. The actual pH of the acetate and PNP buffers must be determined by reference to fresh primary standard phosphate buffer, pH 7.384 at 38°C.

2. *Distilled water.* Free of carbon dioxide, ammonia, bacteria, and mold.

3. *Blood rinse solution.* Bacteria-free parenteral saline [0.85% NaCl (w/v)]. Replace daily.

4. *Electrode wash solution.* Add 1 ml. or 1 g. of concentrated neutral detergent[4] to each 100 ml. of parenteral saline. Replace daily to keep it bacteria-free and mold-free.

5. *Calibrating gases.* Standardized mixtures of CO_2 in air, nitrogen, or oxygen are commercially available, and can be purchased with exact specifications as to desired composition.[5] An assay of CO_2 concentration in percent to the nearest second decimal place (e.g., 4.65% CO_2) should accompany each tank. In addition, however, primary reference gases must be restandardized in one of three ways:

(a) Micro Scholander (16). This is a difficult technique to master and should only be used in a laboratory with extensive experience with this method.

(b) Siggaard-Anderson modified Hempel apparatus (31).

(c) Gas chromatography.

NOTE: The Submitter recommends the following practical approach. Obtain two 80 lb. tanks of precalibrated gas, one tank with about 4% CO_2 in oxygen and the other with about 8% CO_2 in nitrogen. These tanks should be supplied with a percent of volume assay of CO_2 and of oxygen to the nearest second decimal place. The oxygen assay is useful for calibrating a P_{CO_2} electrode. When the tanks arrive, assay the CO_2 concentration by one of the three recommended methods described above. Once the accuracy of the manufacturer's precalibration is checked, keep these two tanks as primary reference standards. For daily use, obtain other precalibrated tanks and check them against the primary reference tanks using the P_{CO_2} electrode. The primary reference tanks will last for several years since they will not be used for routine calibration of the P_{CO_2} electrode. In the past, poor mixing of gases during manufacture necessitated keeping all tanks on their sides and rolling them once a week to prevent layering. However, modern manufacturing insures complete homogeneity of gas mixtures so that rolling is no longer needed. Checker (G. D. B.) has kept one tank standing upright for 4 years without any alteration in concentration.

NOTE: Szönyi (32) recently devised a simple method of determining the CO_2 content of gas mixtures with great accuracy. The gas is equilibrated in a tonometer at 38°C. with a pure 19.91 mM bicarbonate solution. The pH of the

---

[4] Edisonite, Dreft, or Lux Liquid are suitable.
[5] The Matheson Co. Inc., East Rutherford, New Jersey.

solution is measured at 38°C. The pH is subtracted from the factor 9.10206 and the antilog of the resulting number equals the P_{CO_2} of the gas in mm. of Hg when total atmospheric pressure is 760 mm. Hg.

Example: pH = 7.280
            9.10206 − 7.280 = 1.82206
            antilog of 1.82206 = 66.4 mm. Hg

The Submitter has confirmed the validity of Szönyi's method.

6. *Electrode buffer.* Severinghaus (13) used a buffer composed of 0.1 M KCl and 0.01 M NaHCO₃. This buffer was made by adding 7.46 g. of KCl and 0.84 g. of NaHCO₃ to distilled water and bringing to 1 l. The buffer is stable for at least 4 months at room temperature in a polyethylene bottle. An increase in the bicarbonate concentration will decrease the sensitivity of the electrode. Some electrode manufacturers supply a modified buffer solution adapted to the particular requirements of their P_{CO_2} electrodes.

Apparatus

1. *pH meter.* The pH meter should meet these specifications:
(*a*) Sensitivity to at least 0.002 pH units at 37°C.
(*b*) Absolute accuracy to at least ± 0.01 pH units.
(*c*) Temperature compensation accurate to ± 1°C., especially between 35° and 40°C.
(*d*) An input impedance of at least 10^{12} ohms.
(*e*) A buffer adjustment dial with a wide range of compensation. This permits the use of a wide selection of electrodes from different manufacturers while retaining the capacity to read on scale.

2. *pH measuring chain.* The electrode system should have these characteristics:
(*a*) Accurate and simple thermostatic control at body temperature to at least ± 0.2°C.
(*b*) pH measurements on sample volumes less than 100 μl.
(*c*) Shielded glass electrode to minimize capacitance effects.
(*d*) A reproducible junction. Only two commercially available junctions are satisfactory at this time. They are:
  (*i*) The liquid-liquid junction as found in the MacInness-Belcher (9), Sanz (7), and Astrup (8) electrodes. This is a free diffusion static junction formed in a symmetrical narrow-bore tube, and it is very sensitive to blood flow.
  (*ii*) A Leonard palladium junction (33). This is a high velocity,

low-diffusion-volume junction and it is relatively insensitive to blood flow.

NOTE: Fiber, ceramic, fritted-glass, or ground-glass junctions are not satisfactory for blood pH measurements. These junctions are easily fouled and may give false alkaline blood pH results.

NOTE: Checker (W. R. F.) stresses thermostating the junction. Siggaard-Andersen (34) has demonstrated a change in junction potential with a change in temperature. The *junction* must be thermostated, not just the reference electrode.

3. *CO_2 content measurement.* Van Slyke manometric apparatus.[6] Van Slyke volumetric apparatus,[6] AutoAnalyzer,[7] Natelson microgasometer,[8] or Conway diffusion apparatus.[6]

4. *Interpolated P_{CO_2}.* Astrup apparatus.[9]

5. *P_{CO_2} electrode.* Stow-Severinghaus electrode.[10,11,12]

6. *Temperature control.* Circulating water baths with temperature control to at least $\pm$ 0.05°C. are required.[12,13,14]

Sample Collection

COLLECTION SITES

1. *Arterial blood.* The brachial, radial, or femoral artery is used. If multiple samples are required from the same patient, a Cournand needle[15] or a Seldinger catheter,[16] is inserted into the artery using procaine anesthesia and sterile technique. For a single sample the artery is entered with a standard No. 20 or 21 needle. No anesthetic is required if the puncture is made rapidly and accurately.

NOTE: Ideally, arterial punctures should be performed by a skilled physician. In practice, a well-trained and supervised technician may perform this task on adults.

[6] Arthur H. Thomas Co., Vine St. at Third, Philadelphia 5, Pennsylvania.
[7] Technicon Instrument Corp., Chauncey, New York.
[8] Scientific Industries, 220–05 97th Ave., Queens Village 29, New York.
[9] Radiometer AME 1. The London Co., 811 Sharon Dr., Westlake, Ohio.
[10] Spinco Division of Beckman Instruments, Inc., Palo Alto, California.
[11] Radiometer. The London Co., 811 Sharon Dr., Westlake, Ohio.
[12] Instrumentation Laboratories, 9 Galen St., Boston 72, Massachusetts.
[13] Haake. Brinkmann Instruments, Cantiague Rd., Westbury, New York.
[14] Heto. The London Co., 811 Sharon Dr., Westlake, Ohio.
[15] Cournand needle No. 488 LNR. Becton, Dickinson and Co., Rutherford, New Jersey.
[16] Seldinger catheter. Small-size set No. 01-0088. Becton, Dickinson and Co., Rutherford, New Jersey.

2. *Arterialized venous blood.* Heat the hand and forearm in 45°C. water for 5 minutes and draw blood from dilated veins on the back of the hand (35). No tourniquet is needed.

3. *Arterialized capillary blood.* Heat the ear, finger, or heel for 5 minutes with 45°C. water, hot packs, or an electric light bulb. The part can also be rubbed vigorously to stimulate arterial flow. Make a skin puncture at least 5 mm. deep with a scalpel blade.[17] Blood must flow freely and rapidly. If squeezing is required then pH, P_{CO_2}, and oxygen saturation will not be accurate (8, 10, 36, 37, 38, 39, 40, 41, 42, 43).

NOTE: Checker (G. D. B.) uses an Aqua K Pad[18] to obtain safe, thermostatically controlled heating of capillary beds.

4. *Venous blood.* Venous blood can be used for routine clinical evaluation of pH and P_{CO_2} (17, 36, 43, 44, 45, 46, 47, 48, 49, 50). Collect blood with attention to the following details:

(*a*) Place the patient supine for at least 15 minutes and keep him at complete rest.

(*b*) Keep the patient warm, with a blanket if necessary. For best results the skin temperature should be near 35°C. (i.e., warm to touch) (44).

(*c*) Do not permit the patient to "pump" with his clenched fist.

(*d*) Draw blood with or without a tourniquet in place. However, if a tourniquet is used, leave it on while blood is being drawn (36, 51). Use the first aliquot of blood (up to 10 ml.) for pH studies.

COLLECTION VESSELS

1. *Glass syringe.* Use syringes with matched barrels and plungers. Fill the dead space of the syringe and needle with sterile anticoagulant solution. In addition, the plunger can be wet with sterile light mineral oil before the syringe is assembled to insure a perfect seal.

NOTE: Checker (G. D. B.) uses 2 ml. glass syringes for drawing blood from the umbilical artery. He greases the syringes with Vaseline and then autoclaves them. Sterile anticoagulant solution is added to fill the dead space just before blood is drawn.

2. *Plastic syringe.* Plastic syringes have airtight seals and thus do not require oiling. However, some brands develop sticking plungers

[17] A.S.R. SteriSharp No. 11 blade. S.P. No. D 2870-11. Scientific Products, 1210 Leon Place, Evanston, Illinois.
[18] Gorman-Rupp Industries, Bellville, Ohio.

after blood is stored in ice water. It is necessary to test each brand. Plastic syringes are preferable to glass syringes.

3. *Vacuum tube.* Arterial (52) or venous (53, 54) blood can be drawn into heparinized vacuum tubes.[19] Fill the tubes completely. A full 10 ml. vacuum tube has a blood pH only 0.002 to 0.005 units higher than an aliquot drawn into a syringe, whereas a half-filled vacuum tube has a blood pH from 0.02 to 0.04 units higher. The error varies with P_{CO_2}. When P_{CO_2} is high, more carbon dioxide diffuses into the remaining air space causing a greater alkaline error.

4. *Astrup micro tube.* Glass capillary tubes, measuring 1×100 mm. O.D. and having a capacity of 70 μl., are coated with dried heparin (8).[20] Fill the tube completely with arterialized capillary blood avoiding introduction of air bubbles. Insert a small (5×1 mm.) steel wire and seal the ends of the tube with Plasticine. Mix the blood with the dry anticoagulant by stroking a magnet across the length of the capillary, thus moving the wire inside the capillary. The blood in the tube is used for pH measurements and for estimation of P_{CO_2} and bicarbonate with the Astrup tonometer[21]

5. *Natelson micro tube.* This tube measures 4×75 mm. O.D. and has a capacity of 250 μl. One end has a nipplelike tip for easy filling. The tube is available with or without heparin.[22] Fill the tube completely with arterialized capillary blood and insert a short steel wire. Hold the tube between the index finger and the thumb for mixing. Seal the ends with "Critocaps."[23] Measure whole blood pH with a micro glass electrode and whole blood CO_2 content with micro-gasometer. Alternatively, centrifuge the sealed tube in a micro hematocrit centrifuge[24] and measure plasma pH and plasma CO_2 content.

[19] Heparinized Vacutainer tube 10 ml. No. 3400X, 7 ml. No. 3407X or 5 ml. No. 3405X. Becton, Dickinson and Co., Rutherford, New Jersey.

[20] Capillary tubes, wires, sealer, and magnet available from: The London Co., 811 Sharon Dr., Westlake, Ohio.

[21] Micro Astrup Tonometer, Radiometer No. AMT 1. The London Co., 811 Sharon Dr., Westlake, Ohio.

[22] Micro blood collecting tube, short, heparinized SP No. B 3105-2. Scientific Products, 1210 Leon Place, Evanston, Illinois.

[23] Critocap J (large) SP No. B 3110-J and Critocap K (small) SP No. B 3100-K. Scientific Products, 1210 Leon Place, Evanston, Illinois.

[24] Micro hematocrit centrifuge with dual purpose head. Clay-Adams No. CT 2900B (centrifuge) and No. CT 2915 (head). Clay-Adams Inc., 141 E. 25th St., New York 10, New York.

NOTE: The Submitter tested 10 centrifuged micro samples. The tubes were spun 2 minutes and the pH of the separated plasma was never more than 0.01 units higher than the pH of an aliquot centrifuged in a completely filled and sealed 10 ml. test tube.

6. Virtual anaerobic technique of Lilienthal and Riley. Spontaneously formed drops of blood can fall through 1 to 2 cm. of air without detectable change in pH, P$_{CO_2}$, and oxygen saturation (10, 37, 55, 56). Collect falling drops of capillary blood in a small funnel-shaped cup fastened to the end of a tuberculin syringe (glass or plastic) whose dead space is filled with heparin. One to 2 ml. of capillary blood can be collected from adults and 0.4 to 0.8 ml. from small children.

7. Virtual anaerobic technique of Bates and Oliver. Collect capillary blood under sterile mineral oil and then draw into a small glass syringe. The contact time with the oil is very brief and significant loss of CO$_2$ gas into the oil does not occur unless P$_{CO_2}$ is greater than 65 mm. Hg (57).

8. Dilution technique. One part of whole blood or plasma can be diluted with as much as 19 parts of saline (0.9% w/v) without significant change in pH (11, 27). Ware (42) found best results when the saline diluent contained 5 meq./l. of calcium chloride. He collected samples in small polyethylene tubes[25] containing 5 μl. of heparin. Dilutions were made in a syringe.

NOTE: The Submitter has confirmed Ware's results. He collected 100 μl. of whole blood with a standard T.C. micropipet[26] and then he added this blood to 1.0 ml. of diluent in a small plastic cup.[27] A thin layer of light mineral oil was placed on top of the diluent before blood or plasma was added and mixed. The pH of these diluted specimens remained unchanged for 1 hour when refrigerated.

NOTE: There is confusion about the proper and improper use of mineral oil. As indicated above, mineral oil inhibits the free diffusion of CO$_2$ gas from blood to air. When a thin layer of mineral oil is left *undisturbed* overlying a small area of sample, no adverse effects are noted. On the other hand, if a large amount of oil (e.g., 1 ml. of oil plus 5 ml. of blood) is agitated with blood, and if this mixture is centrifuged at high speed, CO$_2$ gas diffuses into the oil (11, 58, 59, 60). The common practice of placing oil in collection tubes for use in hospital wards is not recommended because it is almost impossible to avoid agitation of these tubes after blood is collected.

[25] Polyethylene micro test tube. Spinco Division of Beckman Instruments, Inc., Palo Alto, California.
[26] Accupette. SP No. P 4510B. Scientific Products, 1210 Leon Place, Evanston, Illinois.
[27] Polystyrene cups. Technicon No. 105-394. Technicon Instrument Corp., Chauncey, New York.

SAMPLE STORAGE

If whole blood pH cannot be measured or plasma separated within 20 minutes of withdrawal, then the whole blood should be stored in ice water. Whole blood may be stored in ice water up to 2 hours with less than 0.015 units change in pH. The pH of the average whole blood sample falls from 0.04 to 0.08 units per hour at 37°C., and 0.008 units per hour at 4°C. (61, 62, 63, 64). The actual pH change depends on the glycolysis rate, which in turn is related primarily to the total white count (63). At 25°C. the pH change with time is about half of that at 37°C. Thus, an average fall of 0.005 pH units per 10 minutes can be expected at 25°C.

For precise determination of P_{CO_2} do not store samples in ice water longer than $\frac{1}{2}$ hour (27).

Sodium fluoride has been used to delay glycolysis. The exact concentration is critical. A concentration less than 10^{-4} M fails to inhibit lactic acid production, and a concentration greater than 10^{-2} M causes loss of potassium from cells and alkalinization of blood (32). However, even when the concentration is optimum, sodium fluoride introduces a variable error of from $+0.006$ to -0.014 for pH (63).

SERUM VS. PLASMA VS. WHOLE BLOOD

The pH of serum and heparinized plasma are identical. Clotting has no effect on blood pH (65). The pH of whole blood and plasma are identical when plasma is separated from whole blood at body temperature (3, 66). However, the measured pH of whole blood is about 0.01 units lower than the measured pH of plasma separated at body temperature because of a suspension effect of red cells at the reference electrode (19, 34).

If plasma is separated from red cells at a temperature lower than body temperature, and the plasma pH is subsequently measured at body temperature, the plasma pH will be too alkaline (3, 66, 67).

Whole blood and plasma pH are very temperature dependent because protein ionization varies significantly with small changes in temperature. Since red cells contain more protein than plasma, whole blood pH changes more than plasma pH after any temperature change. Reported average temperature factors for plasma and whole blood are:

Plasma: -0.01 to -0.0118 per °C   (67, 68)
Whole blood: -0.0147 to -0.015 per °C   (67, 69, 70)

Therefore, if whole blood has a pH of 7.40 at 37°C. and is cooled to 4°C., the pH will rise to 7.90. If plasma is separated from this whole blood sample at 4°C. and measured at 4°C., the plasma will also have a pH of 7.90. However, if this cold separated plasma is rewarmed to 37°C. in the electrode, plasma pH will fall back only to 7.56 rather than to 7.40. This false alkaline result occurs because plasma pH changes less than whole blood pH per °C.

These results suggest that plasma pH should not be measured unless plasma can be separated at body temperature. In practice, however, the separation temperature error is constant and small.

NOTE: The Submitter and Checkers (R. E. T. and W. R. F.) prefer to measure plasma rather than whole blood pH to get more stable and reproducible meter readings. Some unstable and unreproducible whole blood pH measurements may be secondary to the high carbonic anhydrase concentration in red cells (71) and to variable red cell suspension effects (34). In addition, separation of plasma permits samples to stand up to 2 hours without change in plasma pH. The Submitter and Checkers (R. E. T. and W. R. F.) accept the relatively constant error of +0.02 to +0.04 pH units to obtain the great convenience and stability of plasma samples.

NOTE: The Submitter found the internal temperature of a large International Centrifuge (Model CS) to range between 31° and 33°C., while the air temperature of the room was 25°C. The highest temperatures were reached when the centrifuge was used just after completing a run. In one experiment a cold centrifuge reached a temperature of 31°C. in 15 seconds. In this same experiment whole blood pH was 7.34 and plasma pH 7.37, an error of +0.03 units; and if the suspension effect of red cells is taken into account the error is only +0.02 pH units.

Blood, when ice cold, should never be centrifuged. Cold samples *must* be rewarmed to room or body temperature before centrifugation.

ANTICOAGULANTS

Heparin has characteristics that make it an ideal anticoagulant (72). In purified form it prevents coagulation of blood in a concentration as low as 1 unit/ml., without altering any known physical characteristics of blood.

NOTE: Use *heparin sodium*, USP, or ammonium heparinate,[28] in a concentration of 1000 units/ml. At one time 1 mg. of heparin equaled 100 units, but this is no longer true since purer preparations are available. USP heparin must now contain at least 120 units/mg. of dry weight, but may contain as high as 150 units/mg.

[28] Ammonium Heparinate. SP No. 51638. Scientific Products, 1210 Leon Place, Evanston, Illinois.

Therefore, heparin doses should be expressed in units—not milligrams. The unitage is based on an *in vitro* assay of anticoagulant activity.

The final heparin concentration should always be less than 100 units/ml. of whole blood. Heparin is supplied in sterile vials at several concentrations. Use the 1000 units/ml. concentration. To heparinize a syringe, draw the heparin solution into the syringe with a sterile needle and coat the inside of the syringe. After removing air bubbles, move the plunger to its most forward position, thus filling the dead space of the syringe and the needle with heparin. The syringe and needle will contain from 100 to 150 units of heparin because the dead space volume varies between 0.1 to 0.15 ml. in a 5 or 10 ml. syringe.

To heparinize capillary tubes, wet the internal surface with the heparin solution. Then dry the tubes in a hot air furnace at a temperature below 110°C. Heparin in the dry state is a suitable anticoagulant for at least 1 year. To heparinize test tubes, wet the internal surface with the heparin solution and dry as above.

Do not use oxalate, citrate, or EDTA for acid-base studies. The oxalate salts cause slight shifts of water and electrolytes between cells and plasma and alter CO_2 content (72). Potassium oxalate produces a slight alkaline shift in blood pH (53). Double oxalate, sodium citrate, and EDTA produce a significant acid shift (53).

I. PROCEDURES FOR pH

General Outline

1. Turn on the meter and water bath at least 15 minutes before making a measurement.

2. Check for proper grounding.

3. Check the cleanliness of all internal and external surfaces.

4. Obtain fresh aliquots of buffer.

5. Remove any distilled water from the electrode surface by replacing it with buffer having the pH nearest to that of blood. Calibrate the meter with this buffer.

6. Wash out the buffer with distilled water.

7. Introduce the buffer whose pH is farthest from that of blood. The meter should read the pH of this buffer to within ±0.005 units if the buffer is about pH 6.8, and to within ±0.02 if the buffer has a pH of about 4.6.

8. Wash the electrode with saline before introducing blood.

9. Introduce whole blood or plasma as a single column free of air bubbles. Never rinse the electrode with blood or plasma.

10. Be certain no minute air bubbles collect at the junction between blood and KCl.

11. Wait 30 seconds after the junction is formed and then take a meter reading. The reading should be stable at this time.

12. Rinse the electrode with saline before reading the next sample.

13. After completing a series of measurements, recheck with the buffer nearest to the pH of blood. This buffer should read within ±0.01 units of its assigned values. If not, repeat the calibration and blood pH measurements.

14. Wash the electrode with detergent.

15. Rinse the electrode with distilled water and leave it filled with distilled water or buffer. If left filled with buffer be certain that the electrode is not permitted to dry out.

Grounding

Incorrect grounding is a frequent cause of inaccurate or unreproducible pH measurements. Absence of true ground, or the presence of too many grounds forming a "ground loop," are equally incorrect.

All grounding should be to *one* common point. Never ground to an electrical outlet *and* to a water pipe. In the United States it is safest to ground to a properly wired three-way outlet with a three-prong plug. If the outlet is not at true electrical ground, then an electrician should correct the situation. Figure 2 shows correct and incorrect grounding arrangements for a meter and circulating water bath.

Cleaning

All internal and external surfaces must be scrupulously clean. The measuring surface of the glass electrode should be cleaned with detergent after each series of blood pH measurements. The electrode is rinsed with saline between samples to prevent euglobulin precipitation and hemolysis of red cells. If protein is deposited on the glass electrode, the response may become sluggish or inaccurate.

NOTE: Sanz (7) recommends cleaning the electrode with pepsin in dilute hydrochloric acid, and Checker (R. E. T.) agrees. However, the Submitter finds neutral detergents equally reliable and easier to use.

The electrode, when not used, must be kept wet with distilled water or acid buffer. Do not permit buffer to dry inside the electrode because this leaves crystalline deposits.

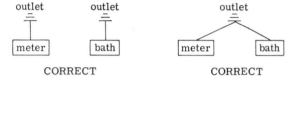

CORRECT   CORRECT

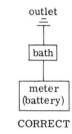

CORRECT

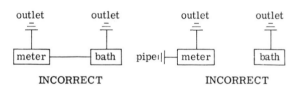

INCORRECT   INCORRECT

FIG. 2. Correct and incorrect grounding of meter and water bath.

The external nonmeasuring surfaces of the glass electrode, the connecting wires, and the reference electrode must be kept *clean* and *dry*. The glass electrode has high electrical resistance. Therefore, all other insulators in the circuit must maintain even higher resistances than glass. Dirt and moisture will permit electrons to leak around the glass electrode to ground, causing falsely low potential readings. In a high humidity environment keep the meter on at all times so as to keep it dry. A room air conditioner or dehumidifier will also help.

Response Check

Two widely spaced buffers (more than 2 pH units) should check within ±0.02 pH units. Perfect response cannot be expected because no glass electrode follows exactly the theoretical pH/millivolt response curve. Most glass electrodes are selected to come within 98% or better of the theoretical slope. The reproducibility of the response is more important than any slight deviation from the theoretically expected response. For example, if an electrode is calibrated with a pH 6.99 buffer and then gives 4.67 with a buffer of known pH 4.65, the electrode should always give 4.67 ±0.005 units upon repeat testing each day. If the electrode system cannot reproduce itself from day to day then either the buffers or electrode system may be faulty. Frequently, the trouble is caused by a dirty glass electrode.

Filling Electrode

Fill the electrode with one continuous column of blood or plasma free of any air bubbles. Never rinse the electrode with blood or plasma. Rinsing with blood or plasma leaves a thin film of plasma sticking to the glass exposed to air. The air exposure will cause false alkaline readings when the next sample is introduced. Always remove all traces of blood or plasma with saline before filling the electrode with a new sample.

Junction Formation

The junction between blood or plasma and saturated KCl is still a "problem area" in pH measurements. The junction potential developed by different buffers and by different blood samples can vary (34, 73). In addition, the junction potential developed between whole blood and KCl is different from that developed between plasma and KCl (19, 34). This difference causes whole blood pH to be about 0.01 units *lower* than that of plasma.

In an attempt to prevent this 0.01 pH difference between whole blood and plasma, Semple (74) and Siggaard-Andersen (34) have tried substituting 0.15 M NaCl for saturated KCl. However, this change radically alters classic blood pH values. Whole blood pH averages 0.1 pH units lower when 0.15 M NaCl is substituted for saturated KCl (34, 74).

Potassium chloride, rather than sodium chloride, has been used in all classic junctions because KCl has equitransferrent ions. If ions are not equitransferrent, then one ionic species will migrate away from the other creating a new potential. Saturated KCl, rather than isotonic KCl, is used because more reproducible junctions are obtained. Potassium nitrate is also equitransferrent. Ten per cent (w/v) potassium nitrate gives the same results with whole blood as saturated potassium chloride (75).

Thermostating

Blood pH, and the potentials at each junction in the measuring chain, change with temperature. Therefore, reproducible blood pH measurements require temperature control to at least ±0.1°C. Best control is obtained with circulating water baths having high flow rates and precision mercury or electronic thermostatic controls.[29,30,31,32]

NOTE: Electrodes without any air space between the heating fluid and the glass electrode have better temperature control than electrodes with an air space. When using an electrode with an air space (e.g., Astrup micro electrode), allow sufficient warm-up time between readings. The Submitter has placed a micro thermistor probe in the air space of an Astrup micro electrode and found a temperature drop of 3°C. after 3 minutes of *continuous* washing and flushing, and it took 3 minutes to return to the original temperature. Decreasing the washing and rinsing time will prevent excessive temperature drops.

Capacitance Effects

Nylon, Orlon, and Dacron uniforms, slips, and lab coats accumulate large capacitance charges. These charges cause erratic meter needle movement when the operator moves. If the operator stands still when measuring pH, then needle fluctuation is minimized. The capacitance effects cause the meter reading to waiver but not to drift. Therefore, capacitance effects do not vary the final reading at rest.

II. PROCEDURES FOR P_{CO_2}

Calculation

1. Measure the pH of whole blood or true plasma.
2. Measure the CO_2 content of plasma in one of the following:
   (a) Van Slyke manometric or volumetric apparatus (72)
   (b) Natelson microgasometer (76)
   (c) AutoAnalyzer (77)

[29] Haake. Brinkmann Instruments, Cantiague Rd., Westbury, New York.
[30] Heto. The London Co., 811 Sharon Dr., Westlake, Ohio.
[31] Instrumentation Laboratories, 9 Galen St., Boston 72, Massachusetts.
[32] Beckman modular blood gas cuvet. Spinco Division of Beckman Instruments, Inc., Palo Alto, California.

(d) Conway diffusion apparatus (78)
(e) Microtitration apparatus (79)

NOTE: Accurate measurement of CO_2 content is important. Collect blood anaerobically and separate the plasma anaerobically. Apply a hematocrit correction (22,72) if whole blood is used in place of serum or plasma.

NOTE: The Submitter has centrifuged whole blood in small capillary tubes[22] with a plastic cap[23] at each end. The tubes were centrifuged in a micro hematocrit centrifuge[24] at top speed for 3 minutes. The pH changed only +0.01 to +0.02 units. The capillary tube was broken at the plasma-red cell interspace and a finger was held over the small end of the tube while the wider end was brought to the pipet tip of the microgasometer. The pipet tip was pressed tightly against the open end of the capillary tube and 30 μl. of plasma was withdrawn into the microgasometer without loss of CO_2. Any other indirect method of transfer tried was always accompanied by some CO_2 loss as evidenced by an alkaline change in plasma pH.

NOTE: The dilution method of measuring pH described above can also be used successfully for the measurement of CO_2 content with the Van Slyke apparatus or with the AutoAnalyzer. Obviously, accurate dilutions are required, and the results obtained must be multiplied by the dilution factor.

3. Refer to one of several available tables or nomograms for estimation of P_{CO_2} from pH and CO_2 content (22, 23, 24, 25). If temperature corrections are not made, calculated P_{CO_2} will have a coefficient of variation of $\pm$ 5% when compared with true P_{CO_2} (27). If proper temperature corrections are made for changes in S and pK (80), then the coefficient of variation is reduced to $\pm$ 2.5% (27).

Direct Measurement

1. Assemble the P_{CO_2} electrode according to the manufacturer's instructions.

2. Check the membrane for leaks by measuring the resistance across the membrane. The resistance should be high.

3. Permit at least 15 minutes warm-up after turning on the water bath.

4. Calibrate the P_{CO_2} electrode with two gases having known but different CO_2 concentrations (usually 4% and 8%). The gases are humidified and warmed to 37°C. just prior to entering the electrode chamber. The actual P_{CO_2} of each gas will vary with barometric pressure and electrode temperature. Calculate the actual P_{CO_2} from the per cent assay as follows:

(a) Read the barometric pressure in the laboratory.

(b) Subtract the saturated water vapor pressure from the total barometric pressure to obtain dry gas pressure.

Temperature	Water vapor pressure
37.0	47.1 mm. Hg
37.5	48.3 mm. Hg
38.0	49.7 mm. Hg

(c) Multiply per cent CO_2 times the dry gas pressure to obtain P_{CO_2}.

Example:

Barometric pressure	= 747 mm. Hg
Water vapor pressure (37°C.)	= 47 mm. Hg
Dry gas pressure	= 700 mm. Hg
P_{CO_2} of 5 00% CO_2	= 5% × 700 = 35 mm. Hg

5. Gases should always flow from the top stopcock down through the chamber and out the lower stopcock. This helps keep the Teflon membrane clean and free of water droplets.

6. Fill the electrode with a fresh sample of whole blood. Introduce the blood through the lower stopcock and inject until blood appears at the upper stopcock. Close the bottom stopcock and leave the upper stopcock open to atmospheric pressure. Do not rinse the electrode with blood or saline before taking a measurement. Keep the electrode flushed with gas just before making a blood measurement. Use a gas whose P_{CO_2} is near that of the sample to be measured.

7. Read P_{CO_2} when equilibrium is reached (usually 2 minutes).

8. Wash the electrode chamber with a saline-detergent mixture and then follow with saline. Prewarm these solutions. Reflush the electrode with CO_2 gas before measuring another blood sample.

NOTE: The Submitter finds it practical to follow one blood specimen with another when measuring a series of samples. He displaces the first specimen with an aliquot of the second. After a short equilibration period, he injects a second aliquot of the second specimen and reads P_{CO_2} at equilibrium. This procedure is not the same as "rinsing" with blood because air is never introduced into the electrode chamber.

Temperature Control

The P_{CO_2} electrode is very temperature sensitive. Temperature must be held closer than ± 0.05°C. If cold blood is injected into the electrode chamber, the equilibration time will be longer. On the other hand, if blood is preheated, P_{CO_2} will increase secondarily to glycolysis.

Calibration

The P_{CO_2} electrode is sluggish. It may take as long as 4 minutes to reach complete equilibrium when P_{CO_2} moves from 0 to 80 mm. Hg. To decrease equilibrium time for blood samples, pre-flush the electrode with gas having a P_{CO} near that of the test sample. The equilibration time varies with:

(1) Membrane thickness. (2) Electrode buffer strength. (3) Electrode buffer volume. (4) Delta P_{CO_2}.

The slope of the calibration curve does not vary as long as the same buffer is used (27).

Precision

The coefficient of variation of direct P_{CO_2} measurements can be as low as ± 0.33% (27). This low coefficient of variation can only be obtained by exercising care and patience during calibration, and by re-checking calibration after each sample.

Samples

Standard electrodes (1964) require as little as 0.2 ml. of whole blood. Electrodes requiring as little as 50 μl. are being developed.

NOTE: Checker (G. D. B.) removes the lower stopcock after gas calibration and pushes the sample meniscus just past the electrode surface to use as little as 0.2 ml. of sample.

P_{CO_2} by Equilibration and Interpolation

1. Collect arterialized capillary blood in a heparinized capillary tube.

2. Measure the actual pH of whole blood.

3. Equilibrate two 50 μl. aliquots of whole blood with two gases of known but differing P_{CO_2}. The carbon dioxide gas should be combined with oxygen because the Astrup technique is based on fully oxygenated blood. Equilibrate for 3 minutes at 38°C. in a micro Astrup tonometer.[21]

4. Measure the pH of the two equilibrated samples.

5. Construct a pH/log P_{CO_2} straight line graph using the points obtained with the equilibrated samples.

6. Estimate the actual P_{CO_2} by locating the actual blood pH on the constructed pH/log P_{CO_2} graph.

7. Estimate bicarbonate concentration from pH and P_{CO_2} with the Siggaard-Andersen nomogram (26).

NOTE: To obtain correct results with the Astrup technique, accurate measurement of blood pH is necessary. Exact values for the P_{CO_2} of the equilibrating gases must also be obtained. In addition, the nomogram is based on studies at 38°C. If blood is obtained from a hypothermic patient, then the nomogram is not valid. Completing an average measurement takes 6 to 8 minutes. If the blood introduced into the pH electrode is blown back into the equilibrating chamber, then as little as 100 μl. of sample is sufficient for determining pH, P_{CO_2}, and bicarbonate.

Normals for pH and P_{CO_2}

Blood pH is very constant in resting adults and there is no sex difference (see tabulation). Brachial vein pH is from 0.00 to 0.03 pH units lower than brachial artery pH (81).

pH IN ADULTS AT REST

Source	No.	Mean	S.D.	Ref.
Venous plasma (brachial)	55	7.40	±0.01	(82)
Venous whole blood (jugular)	50	7.37	±0.015	(46)
Arterial whole blood	15	7.43	±0.02	(83)
	18	7.42	±0.015	(84)
	18	7.41	±0.018	(64)
	50	7.42	±0.016	(46)

NOTE: The venous-arterial pH difference between jugular vein and any artery is greater than the pH difference between the brachial vein and any artery. The greater difference is evidence of greater metabolic activity in brain than in resting muscle (46).

Venous CO_2 content is higher than capillary or arterial CO_2 content. Plasma CO_2 content is higher in men than in women (see tabulation).

PLASMA CO_2 CONTENT IN RESTING ADULT MEN AND WOMEN

Source	No.	Mean (mM./l.)	Range	Ref.
Venous, men	7	30.3	28.6–31.9	(82)
Venous, women	8	27.8	24.9–29.2	(82)
Capillary, men	39	27.5	24.2–31.5	(85)
Capillary, women	17	26.1	23.1–29.4	(85)

P_{CO_2} in venous blood is higher than arterial or capillary blood. P_{CO_2} is higher in men than in women, as shown in the next tabulation.

P_{CO_2} IN RESTING ADULT MEN AND WOMEN

Source	No.	Mean (mm. Hg)	Range	Ref.
Venous, men	7	48.0	44–53	(82)
Venous, women	8	44.5	38–48	(82)
Capillary, men	39	43.9	40–50	(85)
Capillary, women	17	40.1	36–46	(85)
Arterial, men	15	43.7	37–50	(83)

Shock and Hastings (85) studied 56 adult subjects for several months and found no characteristic pattern of fluctuation during day time hours. However, pH falls, P_{CO_2} rises, and CO_2 content falls after exercise (83). After 2 minutes of running, venous pH can fall from 7.37 to 7.13 (61).

pH varies with age. From age 16 to 25 the venous pH ranges from 7.29 to 7.40, while at age 35 the range is 7.35 to 7.40, and at age 70 the range is 7.42 to 7.47 (86).

pH and P_{CO_2} are more variable in newborns and children than in adults (see last tabulation).

pH AND P_{CO_2} OF CAPILLARY BLOOD IN NEWBORNS AND CHILDREN

Age	No.	Mean pH	S.D.	Mean P_{CO_2} (mm. Hg)	S.D.	Ref.
Premature	91	7.31	±0.065	—	—	(87)
Full term, at birth	43	7.34	±0.047	—	—	(87)
Full term, 4 hr. old	10	7.40	±0.042	—	—	(87)
Full term, 8 hr. old	11	7.41	±0.040	—	—	(87)
Full term, 12 hr. old	10	7.44	±0.043	—	—	(87)
Full term, 24 hr old	12	7.43	±0.049	—	—	(87)
1 to 6 yr.	26	7.37	±0.043	37.3	±3.8	(88)
7 to 12 yr.	33	7.40	±0.025	38.0	±2.6	(88)
13 to 17 yr., men	57	7.38	±0.026	41.3	±3.1	(88)

REFERENCES

1. Höber, R., Ueber die Hydroxylionen des Blutes. *Arch. Ges. Physiol.* **99,** 572–593 (1903).
2. Hasselbalch, K. A., and Lundsgaard, C., Elektrometrische Reaktionsbestimmung des Blutes bei Körpertemperatur. *Biochem. Z.* **38,** 77–91 (1912).
3. Parsons, T. R., On the reaction of the blood in the body, *J. Physiol.* **51,** 440–459 (1917).
4. Kerridge, P. T., The use of the glass electrode in biochemistry. *Biochem. J.* **19,** 611–617 (1925).
5. Stadie, W. C., An electron tube potentiometer for the determination of pH with the glass electrode. *J. Biol. Chem.* **83,** 477–492 (1929).
6. Stadie, W. C., O'Brien, H., and Laug, E. P., Determination of the pH of serum at 38 C with the glass electrode and an improved electron tube potentiometer. *J. Biol. Chem.* **91,** 243–269 (1931).
7. Sanz, M. C., Ultramicro methods and standardization of equipment. *Clin. Chem.* **3,** 406–419 (1957).
8. Siggaard-Andersen, O., Engel, K., Jørgensen, K., and Astrup, P., A micro method for determination of pH, carbon dioxide tension, base excess and standard bicarbonate in capillary blood. *Scand. J. Clin. Lab. Invest.* **12,** 172–176 (1960).
9. MacInnes, D. A., and Belcher, D., A durable glass electrode. *Ind. Eng. Chem., Anal. Ed.* **5,** 199–200 (1933).
10. Singer, R. B., Shohl, J., and Bluemle, D. B., Simultaneous determination of pH, CO_2 content, and cell volume in 0.1 ml. aliquots of cutaneous blood. *Clin. Chem.* **1,** 287–316 (1955).
11. Van Slyke, D. D., Weisiger, J. R., and Van Slyke, K. K., Photometric measurement of plasma pH. *J. Biol. Chem.* **179,** 743–761 (1949).
12. Hasselbalch, K. A., Die Berechnung der Wasserstoffzahl des Blutes aus der freien und gebundenen Kohlensäure desselben, und die Sauerstoffbindung des Blutes als Funktion der Wasserstoffzahl. *Biochem. Z.* **78,** 112–144 (1916).
13. Severinghaus, J. W., and Bradley, A. F., Electrodes for blood pO_2 and pCO_2 determination. *J. Appl. Physiol.* **13,** 515–520 (1958).
14. Collier, C. R., Affeldt, J. E., and Farr, A. F., Continuous rapid infrared CO_2 analysis. Fractional sampling and accuracy in determining alveolar CO_2. *J. Lab. Clin. Med.* **45,** 526–539 (1955).
15. Lambertsen, C. J., and Benjamin, J. M., Jr., Breath-by-breath sampling of end-expiratory gas. *J. Appl. Physiol.* **14,** 711–716 (1959).
16. Roughton, F. J. W., and Scholander, P. F., Micro gasometric estimation of the blood gases. I. Oxygen. *J. Biol. Chem.* **148,** 541–550 (1943).
17. Hackney, J. D., Sears, C. H., and Collier, C. R., Estimation of arterial CO_2 tension by rebreathing technique. *J. Appl. Physiol.* **12,** 425–430 (1958).
18. Hughes, W. S., The potential difference between glass and electrolytes in contact with the glass. *J. Am. Chem. Soc.* **44,** 2860–2867 (1922).
19. Severinghaus, J. W., Stupfel, M., and Bradley, A. F., Accuracy of blood pH and pCO_2 determinations. *J. Appl. Physiol.* **9,** 189–196 (1956).
20. Siggaard-Andersen, O., The first dissociation exponent of carbonic acid as a function of pH. *Scand. J. Clin. Lab. Invest.* **14,** 587–597 (1962).

21. Gambino, S. R., The value of pK_1''' in the Henderson-Hasselbalch equation. *Scand. J. Clin. Lab. Invest.* **15**, 104–105 (1963).
22. Singer, R. B., and Hastings, A. B., An improved clinical method for the estimation of disturbances of the acid-base balance of human blood. *Medicine* **27**, 223–242 (1948).
23. Siggaard-Anderson, O., and Engel, K., A new acid-base nomogram. An improved method for the calculation of relevant blood acid-base data. *Scand. J. Clin. Lab. Invest.* **12**, 177–186 (1960).
24. Leitner, M. J., and Thaler, S., A table for calculation of arterial pCO_2 from blood pH and CO_2 content. *Am. J. Clin. Pathol.* **33**, 362–363 (1960).
25. Weisberg, H. F., Revised table of factors (f_p) obtained from pH to calculate plasma (or serum) partial pressure of CO_2 ($P CO_2$) from the total CO_2 content. Modified from "Water, Electrolyte, and Acid-Base Balance," 2nd ed. Williams & Wilkins, Baltimore, Maryland, 1962.
26. Siggaard-Andersen, O., The pH log pCO_2 blood acid-base nomogram revised. *Scand. J. Clin. Lab. Invest.* **14**, 598–604 (1962).
27. Gambino, S. R., Determination of blood pCO_2. *Clin. Chem.* **7**, 236–245 (1961).
28. Bates, R. G., Revised standard values for pH measurements from 0 to 95°C. *J. Res. Nat. Bur. Std. A.* **66a**, 179–184 (1962).
29. Spinner, M. B., and Petersen, G. K., Determining the pH of a phosphate buffer solution for blood measurements. *Scand. J. Clin. Lab. Invest.* **13**, 1–7 (1961).
30. Rossier, P. H., Bühlmann, A. A., and Wiesinger, K., "Respiration-Physiologic Principles and Their Clinical Applications" edited and translated from the German edition by P. C. Luchsinger and K. M. Moser, p. 407. Mosby, St. Louis, Missouri, 1960.
31. Siggaard-Andersen, O., and Jørgensen, K., A gasometric apparatus for direct reading determination of carbon dioxide concentration in gas mixtures. *Scand. J. Clin. Lab. Invest.* **13**, 349–350 (1961).
32. Szönyi, S., Technical points in operating the Astrup apparatus: The determination of the CO_2 content of gas mixtures and the storage of blood for pH determinations. *Clin. Chim. Acta.* **9**, 314–316 (1964).
33. Leonard, J. E., Personal communication. Chief project engineer, electrochemistry, Beckman Instruments, Fullerton, California.
34. Siggaard-Andersen, O., Factors affecting the liquid-junction potential in electrometric blood pH measurement. *Scand. J. Clin. Lab. Invest.* **13**, 205–211 (1961).
35. Goldschmidt, S., and Light, A., A method of obtaining from veins blood similar to arterial blood in gaseous content. *J. Biol. Chem.* **64**, 53–58 (1925).
36. Gambino, S. R., Comparisons of pH in human arterial, venous, and capillary blood. *Am. J. Clin. Pathol.* **32**, 298–300 (1959).
37. Gambino, S. R., Collection of capillary blood for simultaneous determinations of arterial pH, CO_2 content, pCO_2, and oxygen saturation. *Am. J. Clin. Pathol.* **35**, 175–183 (1961).
38. Geubelle, F., and Nicolas-Goldstein, M., Comparison Entre le CO_2 et le pH du Sang Artériel. *Clin. Chim. Acta.* **3**, 480–485 (1958).
39. Maas, A. H. J., and van Heijst, A. N. P., A comparison of the pH of arterial

blood with arterialized blood from the ear-lobe with Astrup's micro glass electrode. *Clin. Chim. Acta.* **6**, 31–33 (1961).

40. Maas, A. H. J., and van Heijst, A. N. P., The accuracy of the micro-determination of the P CO_2 of blood from the ear-lobe. *Clin. Chim. Acta.* **6**, 34–37 (1961).
41. Siggaard-Andersen, O., The acid-base status of the blood. *Scand. J. Clin. Lab. Invest.* **15**, Suppl. 70, 1–134 (1963).
42. Ware, A. G., Nowack, J., and Westover, L., Capillary blood pH by a dilution technic. *Clin. Chem.* **9**, 340–346 (1963).
43. Yoshimura, H., A new micro-glass electrode and the pH of arterial, venous and capillary blood. *J. Biochem. (Japan)* **23**, 335–350 (1936).
44. Brooks, D., and Wynn, V., Use of venous blood for pH and carbon-dioxide studies. Especially in respiratory failure and during anaesthesia. *Lancet* **1**, 227–230 (1959).
45. Gambino, S. R., The clinical value of routine determinations of venous plasma pH in acid-base problems. *Am. J. Clin. Pathol.* **32**, 301–303 (1959).
46. Gibbs, E. L., Lennox, W. G., Nims, L. F., and Gibbs, F. A., Arterial and cerebral venous blood—arterial-venous differences in man. *J. Biol. Chem.* **144**, 325–332 (1942).
47. Paine, E. G., Boutwell, J. H., and Soloff, L. A., The reliability of "arterialized" venous blood for measuring arterial pH and pCO_2. *Am. J. Med. Sci.* **242**, 431–435 (1961).
48. Peters, J. P., Bulger, H. A., Eisenman, A. J., and Lee, C., Total acid-base equilibrium of plasma in health and disease. I. The concentration of acids and bases in normal plasma. *J. Biol. Chem.* **67**, 141–158 (1926).
49. Searcy, R. L., Gordon, G. F., and Simms, N. M., Choice of blood for acid-base studies. *Lancet* **2**, 1232 (1963).
50. Sproule, B. J., Mitchell, J. H., and Miller, W. F., Cardiopulmonary physiological responses to heavy exercise in patients with anemia. *J. Clin. Invest.* **39**, 378–388 (1960).
51. Straumfjord, J. V., Jr., Determination of blood pH. *In* "Standard Methods of Clinical Chemistry" (D. Seligson, ed.), Vol. 2, pp. 107–121. Academic Press, New York, 1958.
52. Leitner, M. J., Simple method of sampling arterial blood. *Am. J. Clin. Pathol.* **40**, 299 (1963).
53. Gambino, S. R., Heparinized vacuum tubes for determination of plasma pH, plasma CO_2 content, and blood oxygen saturation. *Am. J. Clin Pathol.* **32**, 285–293 (1959).
54. Still, G., and Rodman, T., The measurement of the content of carbon dioxide in plasma. *Am. J. Clin. Pathol.* **38**, 435–439 (1962).
55. Lilienthal, J. L., Jr., and Riley, R. L., On the determination of arterial oxygen saturations from samples of "capillary" blood. *J. Clin. Invest.* **23**, 904–906 (1944).
56. Lilienthal, J. L., Jr., and Riley, R. L., On the estimation of arterial carbon dioxide from samples of cutaneous (capillary) blood. *J. Lab. Clin. Med.* **31**, 99–104 (1946).
57. Bates, G. D., and Oliver, T. K., Jr., A micromodification of the bubble

method for direct determination of blood gas tensions. *J. Appl. Physiol.* **17,** 743–745 (1962).

58. Austin, J. H., Cullen, G. E., Hastings, A. B., McLean, F. C., Peters, J. P., and Van Slyke, D. D., Studies of gas and electrolyte equilibrium in blood. I. Technique for collection and analysis of blood, and for its saturation with gas mixtures of known composition. *J. Biol. Chem.* **54,** 121–147 (1922).

59. Gambino, S. R., Mineral oil and carbon dioxide. *Am. J. Clin. Pathol.* **35,** 268–269 (1961).

60. Van Slyke, D. D., and Cullen, G. E., Studies of acidosis. I. The bicarbonate concentration of the blood plasma; its significance, and its determination as a measure of acidosis. *J. Biol. Chem.* **30,** 289–346 (1917).

61. Haugaard, G., and Lundsteen, E., The measurement of pH in blood by means of the glass electrode. *Compt. Rend. Trav. Lab. Carlsberg, Ser. Chim.* **21,** 85–99 (1936).

62. Laug, E. P., A reinvestigation of the phenomenon of a first acid change in whole blood. *J. Biol. Chem.* **106,** 161–171 (1934).

63. Siggaard-Andersen, O., Sampling and storing of blood for determination of acid-base status. *Scand. J. Clin. Lab. Invest.* **13,** 196–204 (1961).

64. Wilson, R. H., pH of whole arterial blood. *J. Lab. Clin. Med.* **37,** 129–132 (1951).

65. Yoshimura, H., Does the pH of the blood change during clotting? *J. Biochem. (Japan)* **22,** 297–302 (1935).

66. Yoshimura, H., and Fujimoto, T., Is the hydrogen gas electrode not applicable to the determination of the pH of oxygenated blood? *J. Biochem. (Japan)* **25,** 493–518 (1937).

67. Rosenthal, T. B., The effect of temperature on the pH of blood and plasma *in vitro. J. Biol. Chem.* **173,** 25–30 (1948).

68. Cullen, G. E., Studies of acidosis. XIX. The colorimetric determination of the hydrogen ion concentration of blood plasma. *J. Biol. Chem.* **52,** 501–515 (1922).

69. Craig, F. A., Lange, K., Oberman, J., and Carson, S., A simple, accurate method of blood pH determinations for clinical use. *Arch. Biochem. Biophys.* **38,** 357–364 (1952).

70. Havard, R. E., and Kerridge, P. T., An immediate acid change in shed blood. *Biochem. J.* **23,** 600–607 (1929).

71. Natelson, S., and Tietz, N., Blood pH measurement with the glass electrode. *Clin. Chem.* **2,** 320–327 (1956).

72. Peters, J. P., and Van Slyke, D. D., "Quantitative Clinical Chemistry," Vol. II: Methods, pp. 57–61. Williams & Wilkins, Baltimore, Maryland, 1932. Reprinted in 1956.

73. Schwab, M., and Wisser, H., Zur Methodik der Blut-pH Messung. *Klin. Wochschr.* **40,** 713–721 (1962).

74. Semple, S. J. G., Observed pH differences of blood and plasma with different bridge solutions. *J. Appl. Physiol.* **16,** 576–577 (1961).

75. Sibbald, P. G., and Leonard, J. E., A study of liquid junction systems in blood pH measurements. *Proc. 1961 San Diego Biomed. Eng. Symp.* Available from Beckman Instruments, Inc., Fullerton, California.

76. Natelson, S., "Microtechniques of Clinical Chemistry," 2nd ed., pp. 152–155. Thomas, Springfield, Illinois, 1961.
77. Skeggs, L. T., An automatic method for the determination of carbon dioxide in blood plasma. *Am. J. Clin. Pathol.* **33**, 181–185 (1960).
78. Conway, E. J., "Microdiffusion Analysis and Volumetric Error," pp. 201–214. Chemical Publ. Co., New York, 1963.
79. Freier, E. F., Clayson, K. J., and Benson, E. S., A system for the analysis of clinical acid-base states based on micro methods for the determination of plasma bicarbonate concentration and blood pH. *Clin. Chim. Acta.* **9**, 348–358 (1964).
80. Severinghaus, J. W., Stupfel, M., and Bradley, A. F., Variations of serum carbonic acid pK' with pH and temperature. *J. Appl. Physiol.* **9**, 197–200 (1956).
81. Gambino, S. R., Choice of blood for acid-base studies. *Lancet* **1**, 726–727 (1964).
82. Gambino, S. R., Normal values for adult human venous plasma pH and CO_2 content. *Am. J. Clin. Pathol.* **32**, 294–297 (1959).
83. Baldwin, E. de F., Cournand, A., and Richards, D. W., Jr., Pulmonary insufficiency. I. Physiological classification, clinical methods of analysis, standard values in normal subjects. *Medicine* **27**, 243–278 (1948).
84. d'Elseaux, F. C., Blackwood, F. C., Palmer, L. E., and Sloman, K. G., Acid-base equilibrium in the normal. *J. Biol. Chem.* **144**, 529–535 (1942).
85. Shock, N. W., and Hastings, A. B., Studies of the acid-base balance of the blood. III. Variation in the acid-base balance of the blood in normal individuals. *J. Biol. Chem.* **104**, 585–600 (1934).
86. Eldahl, A., Brintionkoncentrationen i blodet hos raske mennesker i forskellige aldre og hos patienter med cancer og med hypertoni. *Nord. Med.* **3**, 2938–2940 (1939).
87. Graham, B. D., Wilson, J. L., Tsao, M. U., Baumann, M. L., and Brown, S., Development of neonatal electrolyte homeostasis. *Pediatrics* **8**, 68–78 (1951).
88. Cassels, D., and Morse, M., Arterial blood gases and acid-base balance in normal children. *J. Clin. Invest.* **32**, 824–836 (1953).

PHENYLALANINE*

Submitted by: WILLARD R. FAULKNER, The Cleveland Clinic Foundation, Cleveland, Ohio

Checked by: QUENTIN C. BELLES, Leahi Hospital, Honolulu, Hawaii

MARILYN W. McCAMAN, Institute of Psychiatric Research and Department of Neurology, Indiana University Medical Center, Indianapolis, Indiana

HELMUT J. RICHTER, Ottawa Civic Hospital, Ottawa, Ontario, Canada

ELIZABETH K. SMITH, The Children's Orthopedic Hospital and Medical Center, Seattle, Washington

Introduction

Mental retardation can be effectively prevented in the hereditary metabolic condition, phenylketonuria, if treatment with a diet low in phenylalanine is started early in infancy (1). Consequently, early detection of this condition is vital in giving the full potential benefit of treatment to the patient.

From the time Fölling (2) reported his finding of phenylpyruvic acid in the urine of these patients until recently, presumptive tests for this abnormal metabolite have constituted the standard laboratory diagnosis. The most common of these techniques is the ferric chloride test that results in the formation of a green color when a few drops of ferric chloride are added to a urine specimen (3). Another less common test is based on the production of turbidity upon the addition of 2,4-dinitrophenylhydrazine (4). Although easy to perform, these tests have serious disadvantages, one of which is their lack of specificity for phenylpyruvic acid. Ferric chloride reacts with a number of compounds to form colors or cloudiness or both, thereby causing confusion in the interpretation of results. For example, p-hydroxyphenylpyruvic acid, found in conditions other than phenylketonuria, gives a positive color reaction (3). A simple paper strip test[1] is available which may be more sensitive and more specific than aqueous ferric chloride (5). The 2,4-dinitrophenylhydrazine test, being a gen-

---

* Based on the method of McCaman and Robins (16).

[1] Phenistix, Ames Company, Inc., Elkhart, Indiana.

199

eral one for aldehydes and ketones, gives a positive reaction with acetone and p-hydroxyphenylpyruvic acid as well as with phenyl-pyruvic acid and other carbonyl compounds (3).

The greatest disadvantage of these two tests, as diagnostic procedures, is that phenylpyruvic acid is not constantly excreted in detectable amounts in phenylketonuria. The affected infant frequently does not excrete phenylpyruvic acid until the age of 4 to 6 weeks (6). A negative test at this age would therefore be conducive to dismissing phenylketonuria from consideration as a possible diagnosis, resulting in no treatment at a time when it could be the most beneficial. It should be pointed out that the urine of a phenylketonuric infant may occasionally give a negative test even at an age when it is usually excreted. Therefore, a single negative examination does not constitute sufficient evidence to exclude the disease (7).

In contrast to the delay in appearance of phenylpyruvic acid in the urine, phenylalanine in the blood of a phenylketonuric child rises within a few hours after birth and remains well above the normal range on the first day (7). For this reason, all cases of suspected phenylketonuria should be confirmed by measurement of the serum phenylalanine level. This permits the earliest possible laboratory diagnosis as well as one of the most valid of criteria for evaluating the efficacy of a phenylalanine-deficient diet.

Several methods have been devised for the estimation of serum phenylalanine. This compound may be separated from other amino acids in a protein-free filtrate of plasma by paper chromatography and the amount estimated after reaction with ninhydrin by a visual comparison of the color intensity with that of known amounts of phenylalanine (8). Although used rather widely, this technique allows only the roughest estimation of the actual quantity present.

Several quantitative procedures are based on the Kapeller-Adler test in which phenylalanine is oxidized to benzoic acid, and then nitrated. Subsequently, the nitrated derivative is measured colorimetrically in the presence of hydroxylamine (9, 10). This method is accurate but is not generally applicable to the analysis of micro samples as needed for pediatric patients.

In the method of Udenfriend and Cooper, phenylalanine is enzymatically decarboxylated to phenylethylamine by treatment with Streptococcus faecalis. The phenylethylamine is measured colorimetrically after reaction with methyl orange (11).

The method of La Du, permitting the analysis of 0.1 ml. of serum, is based on the oxidation of L-phenylalanine by snake venom L-amino acid oxidase to phenylpyruvic acid. In the presence of arsenate and borate ions, the phenylpyruvic acid is converted to an enol-borate complex which has a high absorption in the ultraviolet region (12).

Phenylalanine may be separated from other amino acids in a mixture by ion-exchange chromatography and then treated with ninhydrin to yield the characteristic colored reaction product which may be measured spectrophotometrically (13, 14). Although accurate and highly specific, this means of analysis is time-consuming and does not allow the performance of multiple determinations with one piece of apparatus. The apparatus commercially available has the further disadvantage of being relatively expensive.

For the purpose of screening all newborn infants in an institution, Guthrie devised a test to detect elevated phenylalanine levels (15). This test is based on the inhibition of growth of *Bacillus subtilis* by β-2-thienylalanine, and on the ability of phenylalanine as well as other substances (proline, phenylpyruvic acid, and phenyllactic acid) to overcome this inhibition when present in greater than normal quantities in blood. Although this procedure can be of potential value, its usefulness is limited in that the results are not quantitative. Furthermore, the positive specimens must ultimately be confirmed by quantitative serum analyses.

Principle

The procedure described here is that presented by McCaman and Robins (16). It is based on the observation of Lowe, Robins, and Eyerman (17), that fluorescence obtained from the reaction of phenylalanine and ninhydrin is greatly enhanced in the presence of glycyl-DL-phenylalanine. By making certain modifications, this principle was used in the development of a quantitative method for phenylalanine. McCaman and Robins (16), found that the enhancement of fluorescence of the phenylalanine-ninhydrin reaction product by the peptide, L-leucyl-L-alanine, could be made specific in the presence or absence of other amino acids by maintaining the pH at 5.8 ± 0.1. This reaction is highly sensitive for phenylalanine. The method described is simple to perform and requires only 100 μl. of serum for a determination. If a fluorometer with a microcuvet attachment is available, considerably smaller quantities of serum may be used; for example, 10 μl. according to McCaman and Robins (16).

Reagents

1. Succinate buffer, 0.3 M, pH 5.8 ± 0.1. Dissolve 5.94 g. of disodium succinate dihydrate ($NaOOCCH_2CH_2COONa·2H_2O$) in water,
add 4.0 ml. of 1 N HCl, and dilute to 100 ml. Adjust to the correct pH
with either 1.0 N HCl or 1.0 N NaOH using a pH meter. Store at 4
to 10°C.

NOTE: Checker (Q. C. B.) suggests making the buffer by mixing 200 ml. of
0.30 M succinic acid and 340 ml. of 0.30 N NaOH and then adjusting the pH
to 5.8 by adding more of the succinic acid or NaOH as required. An alternate
method of preparation is given by Checker (E. K. S.) as follows: Dissolve 3.54
g. of succinic acid in distilled water. Add saturated (18 N) NaOH to bring the
pH to 5.8.

2. Ninhydrin, 1,2,3-indantrione monohydrate, 30 mM. Dissolve
0.534 g. in water and dilute to 100 ml. Store in a brown bottle at room
temperature.

3. L-Leucyl-L-alanine,[2] *5 mM.* Dissolve 0.1011 g. of L-leucyl-L-
alanine in water and dilute to 100 ml. Dispense in small tubes in
volumes of approximately 1 ml. and store frozen. Discard any remaining thawed reagent.

NOTE: Checker (E. K. S.) reports that this reagent is highly critical and is subject to rapid deterioration even at refrigerator temperatures (4 to 10°C.), and has
therefore made the suggestion together with Checker (M. W. M.) that the peptide
be dispensed in small volumes and stored frozen until the day of use.

4. Copper reagent. Dissolve each compound separately in about
100 ml. of water. Add the solutions in the order listed, mixing after
each addition. Dilute to 1 l.

Sodium carbonate, anhydrous	1.6 g.
Potassium sodium tartrate ($KNaC_4H_4O_6·4H_2O$)	0.100 g.
Copper sulfate ($CuSO_4·5H_2O$)	0.060 g.

This reagent is stable at room temperature for several weeks.

5. Trichloroacetic acid, 0.6 N. Dissolve 98.0 g. of trichloroacetic
acid in water and dilute to 1 l.

6. Trichloroacetic acid, 0.3 N. Dilute 1 vol. of the 0.6 N acid with
1 vol. of water.

7. Phenylalanine stock solution, 100 mg./100 ml. Dissolve 100 mg.
of L-phenylalanine in 0.3 N trichloroacetic acid and dilute to 100 ml.
with the acid. Store at 4 to 10°C.

[2] Obtained from Mann Research Laboratories, 136 Liberty Street, New York 6,
New York.

8. *Phenylalanine working standards.* Dilute the stock standard with 0.3 N trichloroacetic acid according to the tabulation:

Standard No.	Stock (ml.)	Dilute to (ml.)	Phenyl-alanine (mg.)	Equivalent to (mg. phenyl-alanine per 100 ml. serum)
1	5.00	50	10	20
2	4.00	50	8	16
3	2.00	50	4	8
4	1.00	50	2	4
5	1.00	100	1	2
6	0.50	100	0.5	1

Store the working standards at 4 to 10°C.

9. *Succinate-ninhydrin-leucylalanine mixture.* Just before starting a test, pre-mix the succinate buffer, the ninhydrin, and the leucyl-alanine in the volume ratio of 10:4:2. The volume of this mixture needed for each sample is 0.8 ml.

Procedure

1. Deproteinize serum or plasma by adding 0.10 ml. of 0.6 N trichloroacetic acid to 0.10 ml. of specimen. Micro polyethylene centri-fuge tubes[3] or glass 10×75 mm. tubes are suitable for this purpose. Mix thoroughly and allow to stand for 10 minutes. Centrifuge at moderate speed (2000 r.p.m.) for 10 minutes.

2. To 10×100 mm. tubes, add the following:

	Blank (ml.)	Standard (ml.)	Unknown (ml.)
Succinate-ninhydrin-leucylalanine mixture	0.80	0.80	0.80
Trichloroacetic acid, 0.3 N	0.050	0	0
Working standards (each of the 5 concentrations)	0	0.050	0
Supernatant fluid from specimen	0	0	0.050

NOTE: The succinate buffer, ninhydrin, and the leucylalanine may be added separately in the order given and in the quantities, 0.50 ml., 0.20 ml., and 0.10 ml., respectively.

[3] Beckman/Spinco Division, 1117 California Avenue, Palo Alto, California.

204 WILLARD R. FAULKNER

3. Cover the tubes with Parafilm and mix by inversion.

NOTE: Checker (M. W. M.) warns against the use of rubber or cork stoppers, which may introduce fluorescing contaminants into the test solutions. Stoppering was not found necessary in her laboratory.

4. Incubate the tubes for 2 hours in a water bath at 60°C.
5. Cool the tubes in an ice bath. Add 5.0 ml. of the copper reagent to each. Mix thoroughly and allow to stand at room temperature for 10 to 15 minutes.
6. Measure the fluorescence of the standards and the unknown samples, setting the blank at zero fluorescence. If a spectrophoto-fluorometer is available, set the activating (primary) wavelength at 390 mμ and the fluorescing (secondary) wavelength at 485 mμ. If a filter instrument is used, select a filter combination that most nearly reproduces the above wavelengths.

NOTE: The Submitter used Turner instruments, models 110 and 111[4] with a Corning filter No. 7-60 (350 mμ) in the primary and a Wratten No. 8 (sharp-cut filter with starting transmission at 460 mμ) in the secondary. Perhaps the best

TABLE I
FLUOROMETERS AND FILTER COMBINATIONS

Reported by	Instrument	Primary	Secondary
McCaman and Robins (16)	Farrand model A	Corning glass filter No. 5860 (365 mμ)	No. 4303 and 3384 (505 to 530 mμ)
Checker (E. K. S.)	Turner model 110	Corning No. 7-51 (365 mμ)	Wratten No. 8 plus Wratten No. 65A (510 mμ)
Checker (H. J. R.)	Coleman fluorometer, model 12C	Narrow-pass filter with peak transmission at 405 mμ	Sharp-cut filter starting trans-mission at 485 mμ
Checker (Q. C. B.)	Turner model 110	Corning No. 7-60 (350 mμ)	Wratten No. 8 plus Wratten No. 65A (510 mμ)

combination for the secondary is a Wratten No. 8 plus a Wratten No. 65A. The Wratten No. 65A is a narrow-pass filter peaking at 495 mμ. This would eliminate any long wavelength-emitting materials that might be present in a serum specimen (18).

NOTE: Excellent results have been reported with a number of fluorometers using various filter combinations. A few are listed in Table I.

[4] G. K. Turner Associates, 2524 Pulgas Avenue, Palo Alto, California.

7. Plot the fluorescence values of the standards against their corresponding concentrations on rectangular coordinate paper, and draw a curve of "best fit" through the points.

8. Read the values of the unknowns from the calibration curve.

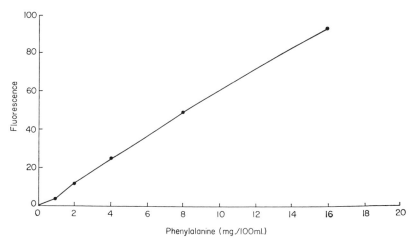

FIG. 1. Typical standard curve of fluorescence–phenylalanine concentration. Method of McCaman and Robins. Turner fluorometer, model 111. Primary filter 110-811(7-60); secondary filter 110-817(8); XI range.

Figure 1 shows a typical standard curve drawn from data obtained in the Submitter's laboratory.

NOTE: The curve is not strictly linear. A sigmoid portion appears in the region of low concentrations, and a slight deviation from linearity is noted in concentrations above 8 mg./100 ml. of phenylalanine. The anomalously reduced fluorescences in the low region are not understood, but were consistently observed with different sets of standards and reagents.

NOTE: Checker (Q. C. B.) reports linearity at least to 40 mg./100 ml. but states that the curve does not pass exactly through zero.

NOTE: Alternatively, the concentration of phenylalanine may be calculated from the formula below, using a standard of approximately the same concentration as the unknown specimen.

$$\begin{array}{c}\text{Concentration of}\\ \text{phenylalanine}\\ \text{in unknown specimen}\end{array} = \begin{array}{c}\text{concentration of}\\ \text{phenylalanine}\\ \text{in standard}\end{array} \times \frac{R_u - R_b}{R_s - R_b}$$

where R_u = fluorescence of unknown specimen; R_b = fluorescence of blank; R_s = fluorescence of standard.

Discussion

Since the level of serum amino acids generally rises after eating, it may be presumed that the level of phenylalanine also rises. However, such an increase would not be sufficiently high to alter the interpretation of a value in the diagnosis of phenylketonuria or the evaluation of a patient's status during therapy.

Analyses were performed on serum rather than plasma in the Submitter's laboratory and on specimens collected without regard to the patient's state of fasting. Plasma may be equally as suitable as serum, but this matter was not investigated.

NOTE: This method was not applied to urine specimens by the Submitter or Checkers. However, Checkers (M. W. M.) states that ammonia would probably constitute the greatest source of difficulty, but that this interfering substance may possibly be removed by a simple resin treatment as described by Saifer and his associates (19).

NOTE: Checker (M. W. M.) reports that serum samples can be stored almost indefinitely at −20°C. with only small losses of phenylalanine. With regard to storage of samples at refrigerator temperatures (5 to 10°C.), the evidence is somewhat conflicting. For example, Checker (Q. C. B.) states that on storage of serum at 5°C., the apparent phenylalanine rises significantly within 5 days and remains high at least up to 14 days. In contrast to this, the Submitter did not find any significant change, either up or down, on storage at 5 to 10°C., up to at least 84 days. Pooled serum samples were stored and analyzed at intervals. The data are presented in Table II.

TABLE II
SERUM PHENYLALANINE CONCENTRATIONS AFTER STORAGE AT 5 TO 10°C.

Specimen	Amount (mg./100 ml.) found after the following days of storage:											Mean	S.D.
	2	3	4	6	7	8	9	42	44	80	84		
1	4.4	5.2	4.0	4.5	4.3	5.3	4.5	5.5	4.5	4.2	4.8	4.7	0.49
2	10.1	11.5	9.5	10.6	10.3	11.5	10.3	10.5	10.3	9.6	10.7	10.4	0.64
3	20.9	22.7	22.8	21.7	—	22.4	—	—	—	21.0	—	21.9	0.84

PRECISION AND ACCURACY OF THE METHOD

To determine the precision of the method, pooled serum specimens at four different levels of phenylalanine concentration were analyzed in replicate. Table III indicates the degree of precision attained in the Submitter's laboratory.

TABLE III
REPRODUCIBILITY OF METHOD

Pooled serum specimen	Replicate analyses	Range	Mean	S.D.
1	11	2.3–3.2	2.8	0.29
2	11	4.0–5.5	4.7	0.49
3	11	9.5–11.5	10.4	0.64
4	6	20.9–22.8	21.9	0.84

As an index to the accuracy of the method, a pooled serum specimen was divided into 6 portions. Nothing was added to the first. Phenylalanine was added to portions No. 2 through No. 6 to increase the initial concentrations by 1.0, 2.0, 4.0, 8.0, and 20.0 mg./100 ml.,

TABLE IV
RECOVERY OF PHENYLALANINE ADDED TO POOLED SERUM

Portion No.	Phenylalanine added (mg./100 ml.)	Phenylalanine found (mg./100 ml.)	Recovery (%)
1	0	2.4	—
2	1.0	3.3	97
3	2.0	4.4	100
4	4.0	6.2	97
5	8.0	10.1	97
6	20.0	20.9	93

respectively. Table IV indicates the absolute and the percentage recoveries obtained.

SERUM PHENYLALANINE CONCENTRATIONS

Table V presents values of serum phenylalanine obtained by the Submitter and by Checkers (M. W. M. and E. K. S.) on individuals in several different categories including phenylketonuric infants.

TABLE V
SERUM PHENYLALANINE CONCENTRATIONS IN HUMAN SERUM

No. of patients	Age range	Type of patient	Reported by	Mean	S.D.
4	Newborn	Normal	Checker (M. W. M.)	1.67	0.77
15	1 mo. to 14 yr.	Normal	Checker (M. W. M.)	1.42	0.58
6	—	Phenyl- ketonuric	Checker (M. W. M.)	33.0	2.04
30	1 mo. to 16 yr.	Normal	Checker (E. K. S.)	1.94	0.53
3	Newborn	Normal	Submitter	1.51	0.82
13	8 mo. to 13 yr.	Normal	Submitter	1.83	0.94
5	9 mo. to 8 yr.	Phenyl- ketonuric	Submitter	26.0	10.0

REFERENCES

1. Knox, W. E., An evaluation of the treatment of phenylketonuria with diets low in phenylalanine. *Pediatrics* **26**, 1–11 (1960).
2. Fölling, A., Über Ausscheidung von Phenylbrenztraubensäure in den Harn als Stoffwechselanomalie in Verbindung mit Imbezillität. *Hoppe-Seyler's Z. physiol. Chem.* **227**, 169–176 (1934).
3. Gibbs, N. K., and Woolf, L. I., Tests for phenylketonuria. Results of a one-year programme for its detection in infancy and among mental defectives. *Brit. Med. J.* **2**, 532–535 (1959).
4. Menkes, J. H., Maple syrup disease. Isolation and identification of organic acids in the urine. *Pediatrics* **23**, 348–353 (1959).
5. Rupe, C. O., and Free, A. H., An improved test for phenylketonuria. *Clin. Chem.* **5**, 405–413 (1959).
6. Hsia, D. Y.-Y., Litwack, M., O'Flynn, M., and Jakovcic, S., Serum phenylalanine and tyrosine levels in the newborn infant. *New Engl. J. Med.* **267**, 1067–1070 (1962).
7. La Du, B. N., The importance of early diagnosis and treatment of phenyl-ketonuria. *Ann. Internal Med.* **51**, 1427–1433 (1959).
8. Berry, H. K., Paper chromatographic method for estimation of phenylalanine. *Proc. Soc. Exptl. Biol. Med.* **95**, 71–73 (1957).
9. Kapeller-Adler, R., Über eine neue Reaktion zur qualitativen and quantitativen Bestimmung des Phenylalanins. *Biochem. Z* **252**, 185–200 (1932).
10. Henry, R. J., Sobel, C., and Chiamori, N., Method for determination of serum phenylalanine with use of the Kapeller-Adler reaction. *A.M.A. J. Diseases Children* **94**, 604–608 (1957).
11. Udenfriend, S., and Cooper, J. R., Assay of L-phenylalanine as phenyl-

ethylamine after enzymatic decarboxylation; application to isotopic studies. *J. Biol. Chem.* **203**, 953–960 (1953).

12. La Du, B. N., and Michael, P. J., An enzymatic spectrophotometric method for the determination of phenylalanine in blood. *J. Lab. Clin. Med.* **55**, 491–496 (1960).

13. Moore, S., Spackman, D. H., and Stein, W. H., Chromatography of amino acids on sulfonated polystyrene resins. An improved system. *Anal. Chem.* **30**, 1185–1190 (1958).

14. Spackman, D. H., Stein, W. H., and Moore, S., Automatic recording apparatus for use in the chromatography of amino acids. *Anal. Chem.* **7**, 1190–1206 (1958).

15. Guthrie, R., Blood screening for phenylketonuria. *J. Am. Med. Assoc.* **178**, 863 (1961).

16. McCaman, M. W., and Robins, E., Fluorimetric method for the determination of phenylalanine in serum. *J. Lab. Clin. Med.* **59**, 885–890 (1962).

17. Lowe, I. P., Robins, E., and Eyerman, G. S., The fluorimetric measurement of glutamic decarboxylase and its distribution in brain. *J. Neurochem.* **3**, 8–18 (1958).

18. Phillips, R. E., Personal communication to the Submitter (1964).

19. Saifer, A., Gerstenfeld, S., and Harris, A. F., Photometric microdetermination of amino acids in biological fluids with the ninhydrin reaction. *Clin. Chim. Acta* **5**, 131–140 (1960).

ALKALINE AND ACID PHOSPHATASE*

Submitted by: Louis Berger, Sigma Chemical Company, St. Louis, Missouri
Guilford G. Rudolph, Vanderbilt University, Nashville, Tennessee
Checked by: George N. Bowers, Jr., Robert B. McComb, and Jean Wenzel, Hartford Hospital, Hartford, Connecticut
John G. Heemstra, Yankton Clinic, Yankton, South Dakota

Introduction

Phosphatase activity in serum and other biological materials is estimated by determining the rate of hydrolysis of various phosphate esters under specified conditions of temperature and hydrogen ion concentration. Glycerophosphate (1, 2, 3) and phenylphosphate (4, 5) have been widely used. Phenolphthalein diphosphate as the sodium salt has been used as a substrate (6, 7). This ester is colorless, while the product, phenolphthalein, is colored at high pH. The phenolphthalein formed is not directly proportional to the enzyme activity (6, 7), while the amount of inorganic phosphate formed is proportional to the enzyme activity. Huggins and Talalay (6) confirmed that the phosphate liberated from sodium β-glycerophosphate and the phenol liberated from disodium monophenylphosphate were proportional to the enzyme concentration. A phosphate ester of p-nitrophenol (p-nitrophenylphosphate) has been used by Ohmori (8) and Fujita (9). The phosphate ester is a colorless compound and the product, p-nitrophenol, at high pH, is colored with an absorption maximum at 400 mμ. Thus, a measure of the product of hydrolysis is a measure of the rate of hydrolysis by the enzyme. The products, p-nitrophenol and phosphate, are formed in direct proportion to the concentration of enzyme (10). This method has been adapted for estimation of total acid phosphatase activity (11) and of prostatic acid phosphatase (12, 13).

Principle

The enzymatic hydrolysis of p-nitrophenylphosphate produces p-nitrophenol and phosphate.

$$p\text{-nitrophenylphosphate} + H_2O \xrightarrow{\text{phosphatase}} p\text{-nitrophenol} + \text{phosphate}$$

* Based on the method of Bessey, Lowry, and Brock (10).

211

The buffered substrate is incubated with serum for 30 minutes. Alkali is added to stop the reaction and to adjust the pH for the determination of the concentration of product formed. The spectral absorbance of p-nitrophenolate (product) has a maximum at 400 mμ, while the substrate (p-nitrophenylphosphate) has a maximum at 310 mμ. The molar absorbance of p-nitrophenolate at 400 mμ is about double that of p-nitrophenylphosphate at 310 mμ. On converting the p-nitrophenolate into p-nitrophenol by acidification, the absorption maximum is shifted to about 320 mμ with no detectable absorption at 400 mμ.

Units

The activities of alkaline and acid phosphatase are expressed in terms of micromoles of substrate transformed per minute per liter (International Units; I. U.), as recommended by the Joint Sub-Commission on Clinical Enzyme Units of the Commission on Clinical Chemistry of the International Union of Pure and Applied Chemistry (14). Alkaline phosphatase activity is expressed as micromoles of p-nitrophenol formed per minute per liter of serum at 37°C. and a pH of 10.2, acid phosphatase as micromoles of p-nitrophenol formed per minute per liter of serum at 37°C. and a pH of 4.7.

The units of activity are calculated by converting the micromoles of product formed per 30 minutes incubation per serum sample to micromoles per minute per liter of serum. The following are examples of calculation for tube 1 of Table I:

International Units alkaline phosphatase activity =

$$\frac{0.03 \ (\mu\text{mole}) \times 1000 \ (\text{factor for conversion to one liter})}{0.1 \ (\text{serum vol.}) \times 30 \ (\text{incubation time}) \times \frac{11.1 \ (\text{vol. of standards})}{11.1 \ (\text{vol. of alkaline determination})}}$$

International Units acid phosphatase activity =

$$\frac{0.03 \ (\mu\text{mole}) \times 1000 \ (\text{factor for conversion to one liter})}{0.2 \ (\text{serum vol.}) \times 30 \ (\text{incubation time}) \times \frac{11.1 \ (\text{vol. of standards})}{6.2 \ (\text{vol. of acid determination})}}$$

Reagents[1]

1. *Alkaline glycine buffer, pH 10.5.* Dissolve 7.5 g. (0.1 mole) of glycine and 95 mg. (0.001 mole) of magnesium chloride ($MgCl_2$) or 203 mg. of $MgCl_2 \cdot 6H_2O$ in 700 to 800 ml. of water; add 85 ml. 1.0 N

[1] The reagents may be obtained from Sigma Chemical Company, 3500 DeKalb Street, St. Louis, Missouri.

(0.085 mole) sodium hydroxide, and dilute to 1 l. The pH should be 10.5 at 25°C. (pH 10.2 at 37°C.). If necessary adjust with 1.0 N NaOH or 1.0 N HCl. Add a few drops of chloroform as a preservative. The buffer is stable about 1 year when stored at 0 to 5°C.

2. *Acid citrate buffer, pH 4.8.*

Solution (a): 0.09 M citric acid. Dissolve 18.91 g. of citric acid ($C_6H_8O_7 \cdot H_2O$) in water and dilute to 1 l.

Solution (b): 0.09 M sodium citrate. Dissolve 26.46 g. of sodium citrate ($Na_3C_6H_5O_7 \cdot 2H_2O$) in water and dilute to 1 l.

Mix 630 ml. of solution (a) with 370 ml. of solution (b). The pH should be 4.8 at 25°C. (pH 4.7 at 37°C.). If necessary, adjust by adding solution (a) to lower the pH, or solution (b) to raise the pH. Add a few drops of chloroform as a preservative. The buffer is stable about 1 year when stored at 0 to 5°C.

3. *Tartaric acid, 0.4 M.* Dissolve 60.0 g. of L(+)-tartaric acid in 400–500 ml. of water. Adjust the pH to 4.8 by adding 2 N NaOH (about 350 ml.); dilute to 1 l. Add a few drops of chloroform as a preservative. The buffer is stable about 1 year when stored at 0 to 5°C.

4. *Tartaric acid buffer, pH 4.8.* Add 100 ml. of 0.4 M tartrate to 900 ml. of 0.09 M citrate buffer (reagent 2). The pH should be 4.8 at 25°C. Add a few drops of chloroform as a preservative. The buffer is stable about 1 year when stored at 0 to 5°C.

5. *Substrate solution.* Dissolve 0.400 g. of disodium p-nitrophenylphosphate tetrahydrate per 100 ml. of water. This solution is 0.012 M and is stable for approximately 8 weeks if kept frozen. As 6.0 μmoles or 2 mg. of substrate are used per determination, it may be convenient to measure 0.5 ml. of the substrate solution into incubation tubes, then stopper and freeze the tubes.

NOTE: p-Nitrophenylphosphate can be synthesized by reacting p-nitrophenol with phosphorus oxychloride in dry pyridine (15). Further purification is effected by recrystallization from cold methanol by addition of acetone and air drying the crystals (16). When stored in well-stoppered containers below 0°C., the colorless air-dried preparations remain stable for at least a year. In aqueous solution hydrolysis to p-nitrophenol and phosphate occurs slowly. This rate is reduced by freezing the solution.

The amount of free p-nitrophenol present in a solution of p-nitrophenylphosphate can be estimated by adding 10.0 ml. of 0.02 N NaOH to 1.0 ml. of substrate solution. Read the absorbance with water as a reference at the wavelength used for the calibration curve. If this absorbance is equivalent to more than 20 I.U. of alkaline phosphatase activity the substrate should be replaced.

6. Stock standard solution. Dissolve 0.0835 g. (0.6 mmole) of *p*-nitrophenol in 6 ml. of 0.10 *N* NaOH. Dilute to approximately 90 ml. with water in a 100 ml. volumetric flask. Add 1.0 *N* HCl dropwise until the solution is pale yellow, then dilute to volume with water. The stock standard is stable for about a year when refrigerated.

7. Working standard solution, 0.06 μmole/ml. Dilute 1.0 ml. of the stock standard solution with 0.02 *N* NaOH to 100 ml. Use the working standard solution only on the day of preparation.

NOTE: *Alternate stock standard solution.*[1] Dissolve 0.1391 g. (1.0 mmole) of *p*-nitrophenol in 10 ml. of 0.10 *N* NaOH. Dilute to approximately 90 ml. with water in a 100 ml. volumetric flask. Add 1.0 *N* HCl dropwise until the solution is pale yellow, then dilute to volume with water. The stock standard is stable for about a year when refrigerated. Prepare a *working standard solution* by diluting 0.60 ml. of the stock standard solution with 0.02 *N* NaOH to 100 ml. This solution has a concentration of 0.06 μmole/ml. Use the working standard solution only on the day of preparation.

8. Sodium hydroxide, 0.02 N and 0.10 N. Prepare from carbonate-free, saturated sodium hydroxide.

9. Hydrochloric acid, concentrated.

Obtaining and Storing Samples

Erythrocytes contain an acid phosphatase (17); therefore, do not use hemolyzed samples of serum for phosphatase determinations. Protect the acid phosphatase activity by keeping the serum in an ice bath (2° to 8°C.), since acid phosphatase is readily inactivated at room temperature and up to 40°C. (18). If the serum is to be kept more than 1 or 2 hours, acidify it by the addition of 0.01 ml. of 20% acetic acid per ml. of serum. Jacobsson (12) has shown that the stability of prostatic acid phosphatase in serum increases with decreasing temperature and decreasing pH, and that at pH 6 there is no decrease in activity for at least 4 hours even at 37°C.

Prostatic manipulation, even by simple rectal palpation, can cause the serum total and prostatic acid phosphatase activities to rise into the abnormal range (13). It is recommended that, if high activities of acid phosphatase are found following rectal examination of the prostate, the acid phosphatase determination be repeated in 24 to 48 hours.

Alkaline phosphatase activity is protected for 1 day by storing the serum at 0° to 5°C., and activity is maintained for 5 or 6 days when the serum is kept frozen at −10° to −15°C. (L. B. and G. G. R.)

Procedures

ALKALINE PHOSPHATASE

1. Transfer 0.50 ml. of alkaline buffer and 0.50 ml. of substrate into each of 2 tubes (reagent blank and sample).
2. Place the tubes in a water bath at 37°C. for 5 minutes (temperature equilibration).
3. Add to one tube 0.10 ml. water (reagent blank). Into the other tube measure 0.10 ml. of the test serum sample. Mix gently and return the tubes to the water bath quickly. Note the exact time the serum was added.
4. Exactly 30 minutes after the addition of serum, add 10.0 ml. of 0.02 N NaOH to that tube. Add 10.0 ml. of 0.02 N NaOH to the reagent blank. Stopper and mix by inversion.
5. Set the spectrophotometer at zero absorbance using the reagent blank as a reference at 400 mμ. Read and record the absorbance of the serum sample.

NOTE: The maximum absorbance of p-nitrophenol is 400 mμ. Any wavelength between 400–420 mμ may be used. The wavelength chosen must be the same for all samples and for the standardization. A wavelength of 410 mμ may be preferred. The absorbance of p-nitrophenol is only slightly less than at 400 mμ, whereas the absorbance of hemoglobin is considerably less at 410 mμ than at 400 mμ. Excess hemolysis must be avoided, since hemoglobin does not absorb the same amount of light in acid solution as it does in alkaline solution.

6. Add 2 drops (about 0.1 ml.) of concentrated HCl to each tube, including the reagent blank, and mix.

NOTE: Acidification shifts the peak absorption from 400 to 320 mμ with no absorption at 400 mμ due to the product, p-nitrophenol. Any absorbance in the tube is due to other absorbing substances in the serum sample, such as hemoglobin or bilirubin. This reading is a serum blank.

7. Read and record absorbance at 400 mμ for the sample using the reagent blank as a reference.
8. Convert the readings in steps 5 and 7 into corresponding units of alkaline phosphatase (from the calibration curve). Subtract the

alkaline phosphatase units of step 7 from the alkaline phosphatase units of step 5. This is the alkaline phosphatase activity of the serum.

NOTE: The ratio of absorbance to concentration of p-nitrophenol is not a constant with all photometers. If the ratio is a constant, the absorbance can be substracted before the units are read from the calibration curve. When the ratio is not a constant, it is necessary to correct for the blank after obtaining the units from the calibration curve.

9. If the alkaline phosphatase activity exceeds the upper limits of the calibration curve, dilute the serum with isotonic saline and repeat the incubation with 0.10 ml. of the diluted serum. Multiply the units of activity by the dilution factor.

TOTAL AND PROSTATIC ACID PHOSPHATASE

If only total acid phosphatase is wanted, use two tubes in step 1 and omit tube c in step 3.

1. Transfer 0.50 ml. of substrate solution into each of three tubes, labeled a, b, and c. Add 0.50 ml. of acid buffer to tubes a and b. Add 0.50 ml. of tartrate acid buffer to tube c.

2. Place the tubes in a water bath at 37°C. for 5 minutes (temperature equilibration).

3. Add 0.20 ml. of serum to tube a (test) and tube c (residual). Mix gently and return them to the water bath quickly. Note the time each serum was added.

4. Exactly 30 minutes after the serum was added, transfer 5.0 ml. of 0.1 N NaOH into all tubes. Add 0.20 ml. serum to tube b (blank).

5. Using water as reference, read and record the absorbance of all tubes at 400 mμ and in the same size cuvets as that used in the standardization.

6. Convert the readings into their corresponding units of acid phosphatase from the calibration curve.

7. Subtract the acid phosphatase units of tube b (blank) from the acid phosphatase units of tube a (test). This is the total acid phosphatase activity of the serum.

8. Subtract the acid phosphatase units of tube c (residual) from the acid phosphatase units of tube a (test). This is the prostatic acid phosphatase activity of the serum.

9. If any reading exceeds the calibration curve, dilute the serum with isotonic saline and repeat the incubation with 0.20 ml. of the diluted serum. Multiply the units of activity by the dilution factor.

Standardization

Prepare a series of tubes as indicated in Table I. Read and record the absorbance of the mixtures at 400 mμ using 0.02 N NaOH as a reference.

NOTE: Any wavelength between 400 and 420 mμ may be used, however, the same wavelength must be used for each determination or standardization.

TABLE I

QUANTITIES USED IN TUBES FOR STANDARDIZATION PROCEDURES

Tube No.	Working standard (ml.)	p-Nitro-phenol (μmole)	0.02 N NaOH (ml.)	Equivalent to I.U. (μmole/min./l.) Phosphatase	
				Alkaline	Acid
1	0.5	0.03	10.6	10	2.8
2	1.0	0.06	10.1	20	5.6
3	2.0	0.12	9.1	40	11.2
4	4.0	0.24	7.1	80	22.4
5	6.0	0.36	5.1	120	33.6
6	8.0	0.48	3.1	160	44.8

Plot absorbance on the ordinate and International Units of acid phosphatase on the abscissa of linear graph paper. Prepare a second curve for International Units of alkaline phosphatase.

Discussion

Two methods for the estimation of serum phosphatase activities are published in "Standard Methods of Clinical Chemistry," Vol. 1 (5) and Vol. 2 (3). The substrates, β-glycerophosphate and phenyl-phosphate for these methods are hydrolyzed in proportion to the phosphatase activity. With β-glycerophosphate, inorganic phosphate is determined before and after incubation. Phosphatase activity is obtained by difference. With phenylphosphate, the product, phenol, is determined. Phenol is determined by a colorimetric method and compared to a stock phenol solution standardized by iodometric titration.

With p-nitrophenylphosphate, the hydrolytic products are formed in proportion to the concentration of enzyme; standardization is with a primary standard, p-nitrophenol; and the measurement of product formed requires a minimum of manipulations.

Phenolphthalein diphosphate has been used as a substrate for alkaline phosphatase (6, 7). The measurement of the product phenolphthalein, requires a minimum of manipulations and standardization is with pure phenolphthalein; however, the formation of phenolphthalein during incubation is not linearly proportional to the concentration of enzyme.

α-Naphthyl acid phosphate has been used as a substrate for serum acid phosphatase (19). The authors claim that the substrate is more specific than other substrates for prostatic acid phosphatase.

With p-nitrophenol as a product, its concentration is determined by reading the absorbance at 400–420 mμ. At these wavelengths bilirubin and hemoglobin absorb light. A wavelength of 410 mμ is preferred over 400 mμ since hemoglobin absorbs less at 410 mμ than at 400 mμ. Hemoglobin absorbs slightly more light in alkaline solution than it does in acid solution; therefore, excess hemolysis should be avoided. For the determination of acid phosphatase hemolysis must be avoided because the erythrocyte contains an acid phosphatase.

Bilirubin has a somewhat different spectrum in alkaline than in acid solution. Bilirubin in alkaline solution has a slightly greater absorbance than bilirubin in acid solution at 400 mμ or 410 mμ; however, the difference is not marked. A serum with 22 mg. bilirubin per 100 ml. had an absorbance difference in alkaline and acid solutions equivalent to 6 I.U. of alkaline phosphatase activity (L. B. and G. G. R.). Correction for this can be made with a serum blank of 0.10 ml. of serum plus 11.0 ml. of 0.02 N NaOH. In the procedure for acid phosphatase a separate serum blank is included because the final dilution of serum is 32, while the final dilution for the alkaline phosphatase determination is 110. Thus, a concentration of bilirubin that would not seriously affect an alkaline phosphatase result could seriously alter an acid phosphatase result. It is essential to have the serum blank in alkaline solution for the acid phosphatase estimation and it may also be desirable for sera of high bilirubin content for alkaline phosphatase estimation.

The inhibition of prostatic acid phosphatase by tartrate was shown by Abul-Fadl and King (20). Inhibition from 0.02 M tartrate, using p-nitrophenylphosphate as substrate, is 95% complete (21). Jacobsson (12) and Ozar, Isaac, and Valk (13) have employed p-nitrophenylphosphate and tartrate to measure serum prostatic acid phosphatase. Fishman and Lerner (22) described a modification of the King-Armstrong method for the determination of total and tartrate-inhibited acid phosphatase. Fishman, Bonner, and Homburger (23) have shown good correlation with prostatic cancer by the determination of tartrate-inhibited phosphatase, while determination of total acid phosphatase had only fair correlation with prostatic cancer. Peterson (24) has concluded that total acid phosphatase activity is a better indication of prostatic cancer than is the determination of the tartrate-inhibited enzyme activity. Ozar, Isaac, and Valk (13) and Jacobsson (12) believe that the determination of serum prostatic acid phosphatase is a better index for the diagnosis of cancer of the prostate than the determination of total acid phosphatase.

Normal values for serum total acid phosphatase activity are between 0.2 and 9.5 International Units (μmole per minute per liter) for women and between 0.5 and 11.0 International Units for men (11). Normal values for serum alkaline phosphatase activity for adults are between 13 and 40 International Units and between 45 and 115 for children (10). An upper limit of normal alkaline phosphatase activity of 60 International Units was found by the Checkers (G. N. B., R. B. M., and J. W.). Jacobsson has reported a normal upper limit of 1.7 I.U. for serum prostatic acid phosphatase activity at pH 5.5 (12). Ozar (13) found a normal upper limit of about 3.0 I.U. for serum prostatic acid phosphatase at pH 4.8. Sewell (25) has shown that the activities of both acid and alkaline phosphatases increase slightly with age through the fifth decade of life with decreases during the sixth and seventh decades.

Elevated values for serum alkaline phosphatase activity are found in cases of abnormal bone metabolism and in hepatic and biliary tract diseases. The proper interpretation of alkaline phosphatase values should include other data relating to calcium and phosphorous metabolism. Hoffman (26) and Cantarow and Trumper (27) have reviewed these subjects. These authors have also reviewed the genetic disorder, hypophosphatasia, in which there is a reduced alkaline phosphatase activity in the serum, leucocytes, and tissues, including bone.

Acid phosphatases are widely distributed, notably in the prostate, stomach, liver, muscle, skin, spleen, and erythrocytes. Acid phosphatase activity in the prostate is at least 100 times that in other tissues. The estimation of serum acid phosphatase is significant mainly in the diagnosis of metastasizing prostatic carcinoma and in connec-

220   LOUIS BERGER AND GUILFORD G. RUDOLPH

tion with the treatment of prostatic cancer. These subjects have been
reviewed by Cantarow and Trumper (27) and by Hoffman (26).

1. Bodansky, A., Phosphatase studies. II. Determination of serum phosphatase.
   Factors influencing the accuracy of the determination. *J. Biol. Chem.* 101,
   93–104 (1933).
2. Shinowara, G. Y., Jones, L. M., and Reinhart, H. L., The estimation of
   serum inorganic phosphate and "acid" and "alkaline" phosphatase activity.
   J. Biol. Chem. 142, 921–933 (1942).
3. Kaser, M. M., and Baker, J., Alkaline and acid phosphatase. *In* "Standard
   Methods of Clinical Chemistry" (D. Seligson, ed.), Vol. 2, pp. 122–131.
   Academic Press, New York, 1958.
4. King, E. J., and Armstrong, A. R., A convenient method for determining
   serum and bile phosphatase activity. *Can. Med. Assoc. J.* 31, 376–381 (1934).
5. Carr, J. J., Alkaline and acid phosphatase. *In* "Standard Methods of Clinical
   Chemistry" (M. Reiner, ed.), Vol. 1, pp. 75–83. Academic Press, New York,
   1953.
6. Huggins, C., and Talalay, P., Sodium phenolphthalein phosphate as a substrate for phosphatase tests. *J. Biol. Chem.* 159, 399–410 (1945).
7. Klein, B., Read, P. A., and Babson, A. L., Rapid method for the quantitative
   determination of serum alkaline phosphatase. *Clin. Chem.* 6, 269–275 (1960).
8. Ohmori, Y., Über die Phosphomonoesterase. *Enzymologia* 4, 217–231 (1937).
9. Fujita, H., Über die Mikrobestimmung der Blutphosphatase. *J. Biochem.*
   (*Tokyo*) 30, 69–87 (1939).
10. Bessey, O. A., Lowry, O. H., and Brock, M. J., A Method for the rapid
    determination of alkaline phosphatase with five cubic millimeters of serum.
    J. Biol. Chem. 164, 321–329 (1946).
11. Andersch, M. A., and Szczypinski, A. J., Use of *p*-nitrophenylphosphate as
    the substrate in determination of serum acid phosphatase. *Am. J. Clin. Pathol.*
    17, 571–574 (1947).
12. Jacobsson, K., The determination of tartrate-inhibited phosphatase in serum.
    Scand. J. Clin. Lab. Invest. 12, 367–380 (1960).
13. Ozar, M. B., Isaac, C. A., and Valk, W. L., Methods for the elimination
    of errors in serum acid phosphatase determinations. *J. Urol.* 74, 150–157
    (1955).
14. King, E. J., and Campbell, D. M., International enzyme units. An attempt at
    international agreement. *Clin. Chim. Acta* 6, 301–306 (1961).
15. Bessey, O. A., and Love, R. H., Preparation and measurement of the
    purity of the phosphatase reagent, disodium *p*-nitrophenyl phosphate. *J.
    Biol. Chem.* 196, 175–178 (1952).
16. Aschaffenburg, R., Preparation of the phosphatase reagent disodium *p*-
    nitrophenyl phosphate. *Science* 117, 611 (1953).
17. Sunderman, F. W., Recent advances in the significance and interpretation
    of phosphatase measurements in disease. *Am. J. Clin. Pathol.* 12, 404–411
    (1942).

18. Woodward, H. Q., A note on the inactivation by heat of acid glycerophosphatase in alkaline solution. *J. Urol.* **65,** 688–690 (1951).
19. Babson, A. L., and Read, P. A., A new assay for prostatic acid phosphatase in serum. *Am. J. Clin. Pathol.* **32,** 88–91 (1959).
20. Abul-Fadl, M. A. M., and King, E. J., Properties of the acid phosphatases of erythrocytes and of the human prostate gland. *Biochem. J.* **45,** 51–60 (1949).
21. Nigam, V. N., Davidson, H. M., and Fishman, W. H., Kinetics of hydrolysis of the orthophosphate monoesters of phenol, *p*-nitrophenol, and glycerol by human prostaic acid phosphatase. *J. Biol. Chem.* **234,** 1550–1554 (1959).
22. Fishman, W. H., and Lerner, F., A method for estimating serum acid phosphatase of prostatic origin. *J. Biol. Chem.* **200,** 89–97 (1953).
23. Fishman, W. H., Bonner, C. D., and Homburger, F., Serum "prostatic" acid phosphatase and cancer of the prostate. *New Engl. J. Med.* **255,** 925–933 (1956).
24. Peterson, C. G., Jr., Prostatic fraction of acid phosphatase versus total acid phosphatase: reliability in diagnosis and treament of prostatic cancer. *J. Urol.* **85,** 643–648 (1961).
25. Sewell, S., Serum acid and alkaline phosphatase values in the adult male. *Am. J. Med. Sci.* **240,** 593–598 (1960).
26. Hoffman, W. S., "The Biochemistry of Clinical Medicine," 3rd ed., pp. 520–525, 551–552. Year Book Medical Publishers, Chicago, Illinois, 1964.
27. Cantarow, A., and Trumper, M., "Clinical Biochemistry," 6th ed., 454–459. Saunders, Philadelphia, Pennsylvania, 1962.

TOTAL PROTEINS IN CEREBROSPINAL FLUID
(COLORIMETRIC) *

Submitted by: HOWARD S. FRIEDMAN, Headquarters, Aerospace Medical Division,
Air Force Systems Command, Brooks Air Force Base, Texas
Checked by: ETHEL CONGER, Grace-New Haven Community Hospital, New
Haven, Connecticut
RUTH D. McNAIR, Providence Hospital, Detroit, Michigan

Introduction

Total proteins in cerebrospinal fluid have been determined colorimetrically by means of the biuret reagent (1, 2), the Nessler reagent (3, 4, 5), the Folin-Ciocalteau phenol reagent (6, 7, 8, 9, 10, 11, 12, 13, 14, 15, 16, 17) and the ninhydrin reagent (18, 19); turbidimetrically with ammonium sulfate (20), sulfosalicylic acid (21, 22, 23), and trichloroacetic acid (24, 25); and immunochemically (26). Except for the last two methods mentioned, all of these procedures require 0.5–2.5 ml. of sample. Because it is difficult to obtain the large volumes of cerebrospinal fluid required by most laboratory procedures, Daughaday, Lowry, Rosebrough, and Fields (7) devised a method using 0.2 ml. of sample. This analytical technique is based on the observations of Herriott (11) and Lowry, Rosebrough, Farr, and Randall (27) that cupric ions markedly enhance the color developed in the presence of the Folin-Ciocalteau phenol reagent by proteins containing tyrosine residues. An ultramicro technique using 0.05 ml. of sample is described by Knights, MacDonald, and Ploompuu (28).

Principle

Proteins in cerebrospinal fluid are treated with alkaline copper tartrate to form cupric-amino acid complexes. The addition of the Folin-Ciocalteau phenol reagent (phosphomolybdotungstic acid) results in the formation of an intense blue color through the reduction of molybdate to molybdenum oxides by both the tyrosine residues and the cupric-amino acid complexes. The color intensity is compared

* Based on the method of Lowry, Rosebrough, Farr, and Randall (27) as modified by Daughaday, Lowry, Rosebrough, and Fields (7).

223

in a spectrophotometer with that developed by a standardized protein solution, and the concentration in the sample is calculated.

Reagents

1. Alkaline carbonate solution. Dissolve 20.0 g. of anhydrous sodium carbonate and 0.50 g. of potassium sodium tartrate (Rochelle salt, $KNaC_4H_4O_6 \cdot 4H_2O$) in 1 l. of 0.10 N NaOH. Store this reagent in a polyethylene bottle. This reagent is stable for at least 6 months.

2. Copper sulfate, 0.10%. Dissolve 1.0 g. of cupric sulfate pentahydrate ($CuSO_4 \cdot 5H_2O$) in 1 l. of water. Store this reagent in a polyethylene bottle. This reagent is stable for at least 3 months.

3. Working alkaline copper reagent. Mix 45 ml. of reagent 1 with 5.0 ml. of reagent 2. Prepare daily.

4. Folin-Ciocalteau reagent. Reflux 100 g. of sodium tungstate dihydrate ($Na_2WO_4 \cdot 2H_2O$), 25.0 g. of sodium molybdate dihydrate ($Na_2MoO_4 \cdot 2H_2O$), 50 ml. of 85% phosphoric acid, 100 ml. of concentrated HCl, and 700 ml. of water gently for 10 hours in an all-glass apparatus. Add 150 g. of lithium sulfate (Li_2SO_4), 50 ml. of water, and a few drops of bromine (or 50 ml. of bromine water). Boil the mixture for 15 minutes without a condenser until the solution is yellow. If a green tint remains, add another drop of bromine (or 10 ml. of bromine water) and boil for another 15 minutes. When cool, dilute to 1 l. with water in a volumetric flask. Dilute the reagent with water so that 1.00 ml. will require 9.0 ± 0.1 ml. of 0.10 N NaOH to neutralize it to phenolphthalein. Store in a dark brown glass or polyethylene bottle. This reagent is stable for 1 year or more.

NOTE: Some commercial reagents[1] are satisfactory, provided that the acidity is adjusted as indicated. To titrate the reagent, transfer 1.00 ml. to a small Erlenmeyer flask, add 10 ml. of water and 1 drop of 0.1% phenolphthalein in 50% ethanol.

To calculate:

$$\frac{\text{ml. for titration} \times 999}{9.0} - 999 = \text{ml. water to add}$$

$$(999 = \text{ml. remaining in the 1 l. volumetric flask})$$

Sample: 10.80 ml. of 0.10 N NaOH are used for the titration.

$$\frac{10.80 \times 999}{9.0} - 999 = 199.8 \text{ ml. of water to add}$$

[1] Standard Scientific Supply Corp., 808 Broadway, New York 3, New York; Hartman-Leddon Co., 60 East Woodland, Philadelphia 43, Pennsylvania.

5. *Standardized pooled serum.* Determine the protein concentration of a centrifuged sample of pooled serum (25–50 ml.) by means of the micro-Kjeldahl method described in Volume 2 of this series.

NOTE: Some commercial protein standards are satisfactory.[2] See the Discussion.

6. *Working protein standard.* Dilute the standardized pooled serum or commercial standard[2] with saturated (0.25%) benzoic acid so that the solution contains an exactly known concentration between 100 and 200 mg. of protein per 100 ml. Store the solution in the refrigerator. This standard is stable for at least 6 months.

NOTE: Checker (E. C.) prefers 0.02% sorbic acid for preservation of diluted serum standards.

Procedure

Place 10.0 ml. of working alkaline copper reagent (reagent 3) in each of three 25 ml. Erlenmeyer flasks. Add 0.10 ml. of cerebrospinal fluid to the first, 0.10 ml. of working protein standard to the second, and 0.10 ml. of water to the third flask for the blank. Mix and let stand at room temperature for 15 minutes. Add rapidly, while swirling the flask, 1.00 ml. of working Folin-Ciocalteau reagent to all mixtures, mix immediately, and let stand at room temperature (25°C.) for 30 minutes or longer. Read the absorbances of the standard and the unknown(s) in a spectrophotometer[3] at 700 mμ, setting the absorbance of the blank to 0. The color is stable for about 15 minutes.

NOTE: Mixing is quite critical. Twenty-five ml. Erlenmeyer flasks and swirling during additions have been found absolutely necessary. Swirling should be vigorous, and it is recommended that each aliquot of phenol reagent be added from a different pipet. Air is blown into the reaction mixture through the pipet to assure rapid and adequate mixing (28).

NOTE: The wavelength of maximum absorbance for molybdenum blue is about 850 mμ. Any wavelength from 520 mμ upwards may be used. The wavelength is not critical because blank, standard, and unknowns are all compared at the same point on the spectral curve. The sensitivity of the reaction is about 3.5 times greater at 700 mμ than at 520 mμ, and about 50% greater at 850 mμ than at 700 mμ. Thus, the sensitivity increases by about 12% for each mμ increment above 520 mμ.

[2] Armour Pharmaceutical Company, Kankakee, Illinois.

[3] The spectrophotometer used for this work by the Submitter was a Coleman Junior, model 6A, with 16 mm. inside diameter cuvets. With 12.5 mm. cuvets, using the Klett-Summerson photoelectric colorimeter, the sensitivity per unit measurement, equivalent to 0.005 absorbance, is approximately 3 mg. per 100 ml.

NOTE: If the absorbance of the unknown is greater than 0.700, beyond which photometric reading errors become excessive, repeat the procedure using a diluted sample of cerebrospinal fluid. Alternately, one of the procedures for serum Total Protein described in Volume 1 of this series may be used for extremely elevated values, with appropriate calculations.

Calculation

$$\text{mg. protein/100 ml. cerebrospinal fluid} = \left(\frac{A_{unk}}{A_{std}} \times C\right) - 6$$

C is the exact concentration of the working protein standard in mg. per 100 ml. See Discussion for comments concerning the value of 6.

Standard Values

The range of values which occurs most frequently in standard medical and clinical chemical texts and references, and which is most often arrived at in independent laboratory studies, is 15–50 mg. total protein per 100 ml. of cerebrospinal fluid. In children, the upper limit of this range may extend to as high as 80 or even 120 mg. per 100 ml. at birth. However, these upper limits gradually decrease to about 50 during the first 5 years of life. For a further discussion of values, see page 235 in this volume.

Discussion

The Folin-Ciocalteau phenol reagent reacts with numerous compounds containing a hydroxyphenyl group. In plasma or cerebrospinal fluid (CSF) this group is found almost entirely in the proteins. A small but quite constant amount of nonprotein color-producing material has been noted (7) in cerebrospinal fluid, equivalent to 6 mg. total protein per 100 ml. For a further discussion of these errors, see page 231 of this volume. The tyrosine residue content of the various proteins in plasma, cerebrospinal fluid, and other tissue fluids varies, within a limited range, under normal physiological conditions. The tyrosine equivalent color of globulins is only slightly more intense than that of albumin (28). Thus, when the ratio of globulins to albumin increases, as in most pathological states, the actual values may be slightly lower than those determined by this method. The determination of the tyrosine content of proteins depends at present on several factors: (1) the completeness of hydrolysis, (2) account of losses during hydrolysis, and (3) specificity and accuracy of the method used. Even the mildest hydrolysis, e.g., with barium hydroxide, tends to produce losses up to 10% (29). This figure is difficult to determine, due to variables involved in different techniques. The Folin-Ciocalteau phenol reagent is known to react with tryptophan, as well as with various drugs and other substances (28). Thus, published values for tyrosine in various proteins are generally accepted to be correct within ±10%. Bovine serum albumin has an estimated tyrosine content of $5.6 \pm 0.56\%$, or 5.04–6.16%. Human serum albumin has an estimated tyrosine content of $5.5 \pm 0.55\%$, or 4.95–6.05%, which considerably overlaps the range for bovine albumin. Increased spinal protein values are almost invariably due to increased spinal fluid globulins, resulting either from tissue antibody formation, decreased selectivity of the blood-brain barrier, or both. Thus, the only absolute standard for determining spinal fluid protein by any method which estimates a residue, the concentration of which varies among the individual protein species which comprise the "total," is a series of standards of varying proportions of albumin and the globulins, each being used at the corresponding ratio and concentration of the spinal fluid protein in question. This is patently impractical and impossible. Therefore, the best approach is to use either a standardized pooled serum containing the same protein species as spinal fluid, or bovine albumin, which is easily obtained in a very pure and accurately standardized form. While the use of a pooled CSF standardized by the micro-Kjeldahl method is theoretically preferable to that of a pooled serum, the protein N content of normal CSF is 2–8 mg./100 ml., while the NPN content is 20–30 mg./100 ml. resulting in additional errors of both determination and calculation.

A method based on the use of the biuret reagent (2) is more precise than that of the copper-phenol reagent because it depends entirely upon the number of peptide bonds ($-CONH-$) present in the proteins. The percentage of peptide bonds varies only slightly from one protein to another, depending upon the type and number of amino acid residues. Although the type and number of amino acid residues may vary considerably from one protein to another, the number of peptide bonds per unit weight of protein is fairly constant, considerably more so than that of tyrosine residues. However, even under optimum conditions the biuret reaction is about 50 times less sensitive than the modified Folin-Ciocalteau reaction.

In common with other heteropoly acid reagents, the exact mechanism of the reaction of the Folin-Ciocalteau reagent with tyrosine residues and cupric-amino acid complexes is still in doubt. If the latter complexes involve free amino groups, then it would seem probable that partial hydrolysis of the proteins would intensify the color per unit weight of protein.

NOTE: There is disagreement on this point between the Submitter and the Checkers. However, it is worthy of further investigation provided that partial hydrolysis, e.g., brought about by heating the initial reaction mixture before color development to 56°C. or higher, will offer the advantages of shortening the time required for the analysis and increasing the color intensity sufficiently to warrant modification of the present method.

Cupric ion concentrations greater than 0.006%, w/v, gave maximum color intensity with the Folin-Ciocalteau reagent under the conditions described above. The cupric ion concentration of the working alkaline copper reagent is 0.01%, w/v.

According to Knights, MacDonald, and Ploompuu (28), the Folin-Ciocalteau reagent is only active for a few seconds at pH 10, which is optimal for the color development. For this reason special emphasis is placed on the rapidity of mixing the reagent aliquot with the initial reaction mixture.

With the use of 0.10 ml. of sample in the procedure, the sensitivity of the method is approximately 2 mg./100 ml. for an absorbance of 0.005 when using 16 mm. cuvets.[3] This can easily be reduced to 1 mg./100 ml. by using 0.20 ml. of sample, but the useful upper limit of the procedure, viz., 280 mg./100 ml. at 0.700 absorbance, is thereby reduced by one half. Absorbance readings within the standard range of 15–50 mg./100 ml. generally fall between 0.040 and 0.125. Since the photometric reading error in this range is about 0.005, the determinative error will be approximately 4–12% in the standard range.

Knights, MacDonald, and Ploompuu (28) report that a large variety of drugs, including salicylates, chlorpromazine, chloromycetin, and penicillin, interfere with this procedure by giving erroneously high values. In such cases, each result must be evaluated on the basis of past medical history and other pertinent factors.

NOTE: If necessary, simultaneous analysis of the cerebrospinal fluid and of its protein-free filtrate (e.g., from equal volumes of cerebrospinal fluid and 5% trichloroacetic acid), may be performed. Correction of the result determined in the untreated sample is then possible.

Values obtained within the range of 15–50 mg./100 ml. may be considered as normal. Extremely high values, i.e., more than twice the upper limit of the range, or above 100 mg./100 ml. may be considered pathological, provided that drug interference can be ruled out. Values between 50 and 100 mg./100 ml., forming the bulk of clinical results, must be carefully screened and evaluated.

REFERENCES

1. Dittebrandt, M., Application of the Weichselbaum biuret reagent to the determination of spinal fluid protein. *Am. J. Clin. Pathol.* 18, 439–441 (1948).
2. Weichselbaum, T. E., An accurate and rapid method for the determination of protein in small amounts of blood serum and plasma. *Am. J. Clin. Pathol.* 10, 40–44 (1946).
3. Bernhard, A., The determination of total protein, sugar and chloride in cerebrospinal fluid. *J. Lab. Clin. Med.* 23, 179–180 (1937).
4. Cipriani, A., and Brophy, D., A method for determining cerebrospinal fluid protein by the photoelectric colorimeter. *J. Lab. Clin. Med.* 28, 1269–1272 (1943).
5. Hubbard, R. S., and Garbutt, H. R., The determination of protein in cerebrospinal fluid. *Am. J. Clin. Pathol.* 5, 433–442 (1935).
6. Andersch, M., and Gibson, R. B. The colorimetric determination of plasma proteins. *J. Lab. Clin. Med.* 18, 816–820 (1933).
7. Daughaday, W. H., Lowry, O. H., Rosebrough, N. J., and Fields, W. S., Determination of cerebrospinal fluid protein with the Folin phenol reagent. *J. Lab. Clin. Med.* 39, 663–665 (1952).
8. Folin, O., and Ciocalteau, V., On tyrosine and tryptophane determinations in proteins. *J. Biol. Chem.* 73, 627–650 (1927).
9. Greenberg, D. M., The colorimetric determination of the serum proteins. *J. Biol. Chem.* 82, 545–550 (1929).
10. Greenberg, D. M., and Miralubova, T. N., Modification in the colorimetric determination of the plasma proteins by the Folin phenol reagent. *J. Lab. Clin. Med.* 21, 431–435 (1936).
11. Herriott, R. M., Reaction of Folin's reagent with proteins and biuret compounds in the presence of cupric ions. *Proc. Soc. Eptl. Biol. Med.* 46, 642–644 (1941).
12. Ling, S. M. The determination of protein in spinal fluid with a note on the increased protein in the spinal fluid in typhus fever. *J. Biol. Chem.* 69, 397–401 (1926).
13. Matz, P. B., and Novick, H., Estimation of protein in cerebrospinal fluid. *J. Lab. Clin. Med.* 15, 370–385 (1930).
14. Rappaport, F., and Lasowski, E., A method for exact protein determination in small quantities of spinal fluid. *J. Lab. Clin. Med.* 28, 1640–1642 (1943).
15. Salt, H. B., Microphotometric determination of globulin and total protein in cerebrospinal fluid or diluted blood serum. *J. Lab. Clin. Med.* 35, 976–982 (1950).
16. Walker, B. S., and Bakst, H. J., Protein analysis in cerebrospinal fluid. A comparative study of methods. *J. Lab. Clin. Med.* 20, 312–314 (1934).

17. Wu, H., and Ling, S. M., A colorimetric determination of protein in plasma, cerebrospinal fluid and urine. *Chinese J. Physiol.* **1**, 161–168 (1927).
18. Saifer, A., Estimation of increased gamma globulin and fibrinogen in cerebrospinal fluid; a serial dilution flocculation method. *J. Lab. Clin. Med.* **36**, 130–133 (1950).
19. Saifer, A., and Norby, T., The photometric microdetermination of gamma globulin in cerebrospinal fluid by a quantitative protein flocculation-ninhydrin reaction. *J. Lab. Clin. Med.* **42**, 316–325 (1953).
20. Looney, J. M., and Walsh, A. I., The determination of spinal fluid protein with the photoelectric colorimeter. *J. Biol. Chem.* **127**, 117–121 (1939).
21. Ayers, J. B., Dailey, M. E., and Fremont-Smith, F., Denis-Ayers method for the quantitative estimaion of protein in the cerebrospinal fluid. *Arch. Neurol. Psychiat.* **26**, 1038–1042 (1931).
22. Ayers, J. B., and Foster, H. E., Quantitative estimation of the total protein in cerebrospinal fluid. *J. Am. Med. Assoc.* **77**, 365–369 (1921).
23. Denis, W., and Ayers, J. B., A method of quantitative determination of protein in spinal fluid. *Arch. Internal Med.* **26**, 436–442 (1920).
24. Meulemans, O., Determination of total protein in spinal fluid with sulfosalicylic acid and trichloroacetic acid. *Clin. Chim. Acta* **5**, 757–761 (1960).
25. Rice, E. W., and Loftis, J. W., Critique of the determination of proteins in cerebrospinal fluid: Evaluation of the biuret method of Goa and the TCA-turbidimetric method of Meulemans. *Clin. Chem.* **8**, 56–61 (1962).
26. Kabat, E. A., Glusman, M., and Knaub, V., Quantitative estimation of albumin and gamma globulin in normal and pathological cerebrospinal fluid by immunochemical methods. *Am. J. Med.* **4**, 653–662 (1948).
27. Lowry, O. H., Rosebrough, N. J., Farr, A. L., and Randall, R. J., Protein measurement with the Folin phenol reagent. *J. Biol. Chem.* **193**, 265–275 (1951).
28. Knights, E. M., Jr., MacDonald, R. P., and Ploompuu, Jr., "Ultramicro Methods for Clinical Laboratories," 2nd ed., pp. 151–154. Grune & Stratton, New York, 1962.
29. Olcott, H. S., Methods for the determination of amino acids. *In* "Amino Acids and Proteins" (D. M. Greenberg, ed.), pp. 92, 105. Thomas, Springfield, Illinois, 1951.

TOTAL PROTEINS IN CEREBROSPINAL FLUID (TURBIDIMETRIC) *

Submitted by: EUGENE W. RICE, William H. Singer Memorial Research Laboratory, Allegheny General Hospital, Pittsburgh, Pennsylvania
Checked by: THEODORE PETERS, JR., Mary Imogene Bassett Hospital, Coopers-
town, New York
ROBERT G. SCHOENFELD, Veterans Administration Hospital, Albu-
querque, New Mexico
MARGARET SMITH, Good Samaritan Hospital, Portland, Oregon

Introduction

At present there is no highly accurate and precise procedure available for the determination of total proteins in cerebrospinal fluid
(CSF) ideally suitable for *routine* use in a hospital clinical laboratory. Methods currently employed vary greatly both in accuracy and
in complexity. Results obtained by Kjeldahl-nitrogen analysis are
generally accepted as "ultimate" reference values (1, 2, 3, 4, 5, 6).
Kjeldahl procedures require up to 15 ml. of sample (3), are difficult to perform (3), and are time consuming, as long as 2 to 3
hours being required for acid digestion (3, 7).
Numerous attempts have been made to develop an accurate, simple,
and practical technique. The tyrosine-equivalence method of Johnston-
Gibson has been employed as a "reference" since it gives results comparable to those of Kjeldahl methods (1, 2). It is less involved than
the Kjeldahl, but more complex than certain other procedures. The
sensitivity of the original Folin-Ciocalteau phenol reagent used by
Johnston-Gibson is greatly increased by the addition of cupric ions.
Daughaday, Lowry, Rosebrough, and Fields adapted a copper-phenol
reagent to the *direct* measurement of CSF proteins, i.e., without preliminary precipitation of proteins (8) (see preceding article by
Friedman). These investigators subtracted an average fixed correction
of 6 mg. protein/100 ml. from all protein values to compensate for
color-producing nonprotein substances. However, Sevensmark (9)
showed that errors as large as 39 mg./100 ml. could occur despite
the use of Daughaday's correction. Sevensmark found it necessary

* Based on the method of Meulemans (13).

231

to determine individual correction values to avoid errors, thereby complicating the procedure appreciably. Zondag and van Boetzelaer (10) and Rieder (11) have also cautioned against employing tyrosine-equivalence methods for the direct measurement of CSF proteins, because many common drugs interfere seriously with the color reaction. However, the high sensitivity of tyrosine procedures permits the analysis of small volumes of CSF, thereby making the Daughaday method particularly useful in pediatrics.

Numerous biuret methods have been applied to the determination of proteins in CSF. Most give consistently higher values than Kjeldahl analyses (12). Results obtained with the biuret method devised by Goa, however, compare quite favorably with the Johnston-Gibson method (12).

Meulemans studied the turbidimetric measurement of proteins in CSF (13) and, in agreement with previous investigators (2, 14, 15), reported that turbidimetric methods employing sulfosalicylic acid do not give accurate results. The values depend markedly upon the relative concentrations of albumin and globulins, since the degree of turbidity with albumin is greater than with globulins. Meulemans described two improved turbidimetric procedures. This chapter describes the simpler of these two techniques (13).

Principle

Proteins in CSF are precipitated by dilute trichloroacetic acid solution and the turbidity of the resulting uniform suspension is measured spectrophotometrically at a wavelength of 450 mμ.

Reagents

1. *Trichloroacetic acid solution, 3.0% (w/v)*. Store in a refrigerator.
2. *Sodium chloride solution, 0.90% (w/v)*. Store in a refrigerator.
3. *Protein standard solution, Armour.*[1] This secondary standard solution is composed of crystalline bovine albumin which has been accurately analyzed for its nitrogen content. It is supplied in an ampule containing 3 ml. of a sterile stable solution with an average nitrogen value of 10 mg. N/ml., the actual content varying from lot to lot. Since

[1] Armour protein standard solution is available as catalog number 3200 from Armour Pharmaceutical Company, Kankakee, Illinois, and also from Aloe Scientific, 1831 Olive Street, St. Louis 3, Missouri, as catalog number VX-27470. Checker (T. P.) found the Armour standard solution to be within 1% of its stated value.

bovine albumin is approximately 16.0% nitrogen, the conversion of mg. N/ml. of this standard to g. albumin/100 ml. is given by the formula: g. albumin/100 ml. = mg. N/ml ÷ 1.60.

NOTE: If desired, a standard protein solution may be prepared from crystalline human or bovine albumin and standardized by its absorbance at 279 mμ (16), by nitrogen determination, or by careful application of a biuret procedure. Absorbance at 279 mμ must be determined before adding benzoic acid (T. P.).

4. Protein standard solution, routine working standard. Prepare a standard solution containing 25–30 mg. albumin/100 ml. by accurately diluting a newly opened ampule of the Armour standard with 0.90% sodium chloride. Saturate the working standard with approximately 0.3 g. of benzoic acid/100 ml. and store in a refrigerator. It is stable for at least several weeks.

NOTE: Additional standard solutions of varying concentrations may be prepared by appropriately diluting the Armour standard with saline. Each laboratory must establish for itself the upper limit of linearity with the particular photometer used. Once the concentration-absorbance relationship is known, only one standard (25–30 mg. albumin/100 ml.) need be included with each series of unknowns.[2]

Procedure

1. To a test tube or cuvet containing 1 vol. of blood-free, centrifuged CSF, add dropwise, with swirling, exactly 4 vol. of trichloroacetic acid solution.

NOTE: The actual volumes of solutions used depend on the minimum capacity of the cuvet chosen.

2. Mix immediately by tapping or inverting the container, and allow to stand at room temperature (22–25°C.) for 10 minutes. A routine working standard solution is treated similarly and included in each set of CSF samples.

[2] Checker (T. P.) found the standard curve to be linear to at least 125 mg./100 ml. in a Bausch and Lomb Spectronic 20 spectrophotometer (19 mm. round cuvets) and in a Klett colorimeter. Checker (R. G. S.) obtained a straight line curve up to 100 mg./100 ml. using 1.0 cm. cuvets with a Beckman B spectrophotometer. The absorbance at this upper level was 0.95. Occasionally, protein concentrations below 20 mg./100 ml. did not fall exactly on the curve, but above 30 mg./100 ml. the curve was always a straight line. Checker (M. S.) routinely used a Coleman Universal spectrophotometer, Coleman cuvets and a total volume of 5 ml., but also found the 10 × 75 mm. micro Coleman cuvets saisfactory for high CSF protein concentrations, though absorbances were too small for "low normal" values.

NOTE: Checker (T. P.) noted a strong temperature effect in the turbidity produced, with a sharp increase in turbidity above 25°C. Hence, if "room temperature" is greater than this, all tubes should be cooled to the indicated temperature range by immersing in a beaker of water.

3. At the end of this 10-minute period, re-mix the contents of all tubes, and determine the absorbances *promptly* against a "saline-reagent" blank (consisting of a mixture of 1 vol. of saline and 4 vol. of trichloroacetic acid) at a wavelength of 450 mμ or with a blue filter (Klett 42 or 45).

NOTE: The 10-minute interval is important. (M. S.) found that after 20 minutes unknowns increase appreciably in turbidity, whereas standard solutions change very little. It is convenient to keep a "saline-reagent" blank in a tightly stoppered cuvet which should be replaced weekly.

Calculation

$$\text{mg. protein/100 ml. CSF} = \frac{A_{CSF}}{A_{std}} \times \text{mg. albumin/100 ml. in standard}$$

NOTE: If the value of the unknown exceeds the established upper limit, repeat the determination with an appropriate saline dilution of the CSF sample. Correct the final value according to the dilution employed.

Discussion

Rice and Loftis (12) have evaluated the Meulemans procedure by comparing 50 CSF samples with values obtained by the Goa biuret method. The differences ranged from −5 to +5 mg. protein/100 ml. with a mean difference of +0.5 mg/100 ml. These investigators also analyzed a series of 24 CSF samples by three methods (Goa, Johnston-Gibson, and Meulemans), and obtained very similar results. The mean difference between the Johnston-Gibson method and the trichloroacetic acid turbidity procedure was +0.7 mg. protein/100 ml.

NOTE: Checker (M. S.) compared the present method with the direct method of Daughaday (8). She included 50 CSF specimens (approximately 15–1075 mg. protein/100 ml.) in the study, used 1 ml. serological pipets for measuring both samples and for diluting with saline, and established concentration-absorbance relationships with the Coleman Universal and the Beckman DU spectrophotometers. The results were in excellent agreement, and the values were distributed randomly. The turbidimetric method gave significantly higher values with three out of four xanthochromic samples. (This error could probably be reduced appreciably by reading xanthochromic specimens against blanks composed of 1 vol. of unknown and 4 vol. of saline.) This study indicates a more favorable evaluation of Daughaday's direct method than those previously cited in the Introduction (9, 10, 11).

The present turbidimetric procedure is simple and rapid. Replicate samples show precision within ±1%. Values are accurate to ±5–10% for various humans and bovine albumin preparations, diluted sera, and spinal fluids in the clinically important range. The expected accuracy is within ±5–6 mg. protein/100 ml. (T. P.). The contrast medium "Pantopaque" interferes with the method (M.S.).

Interpretations (17)

The "normal" range of total proteins in CSF must be considered in relation to the age of the subject. Newborn infants have a high CSF protein content (60–90 mg./100 ml.), presumably because of a poorly developed blood-brain barrier. The protein concentration of fluid obtained from children and adults by lumbar puncture is about 15–45 mg./100 ml. Cisternal fluid contains about 25 mg./100 ml. The level in ventricular fluid is appreciably lower, usually about 10 mg./100 ml. There is a tendency toward higher protein levels in elderly adults, with occasional values up to 60 mg./100 ml. in people over 65 years old. The proteins normally present consist of about 80% albumin. The most common abnormality in CSF is an increase in total proteins. Although in all cases albumin remains predominant, qualitative tests for globulins are more easily performed, and hence used most commonly to indicate increases in total proteins. Fibrinogen is occasionally present in CSF when there is a considerable increase in proteins, and may give rise to a clot on standing.

Proteins are increased in a variety of pathological conditions involving the meninges, brain, and cord. Large increases occur with severe hemorrhages into the CSF. Aside from hemorrhage, meningitis causes the highest concentration of proteins in the CSF. In meningococcic and other pyrogenic meningitides, the exudation from the inflamed meninges is great enough to make the fluid cloudy with many leucocytes and bacteria. The proteins may range from 500–2000 mg./100 ml.

Elevated concentrations of proteins in CSF (50–500 mg./100 ml.) are seen in the nonpurulent meningitides, such as tuberculous and syphilitic meningitis, and aseptic meningeal reaction. A moderate elevation of proteins is found in various involvements of the neural tissue of the brain or spinal cord, as in encephalitis, poliomyelitis, central nervous system syphilis, central paresis, and in a block of the spinal canal by a tumor or a fracture dislocation of the vertebrae.

REFERENCES

1. Wickoff, H. S., and Kazdon, P., A critique of methods for determination of protein in cerebrospinal fluid. *Am. J. Clin. Pathol.* **21**, 1173-1177 (1951).
2. Sethna, I., and Tsao, M. U., Protein level in cerebrospinal fluid: An evaluation of some methods of determination. *Clin. Chem.* **3**, 249–256 (1957).
3. Tourtellotte, W. W., Parker, J. A., Alvin, R. E. and DeJong, R. N., Deternitrogen of toal protein in cerebrospinal fluid by an ultramicro-Kjeldahl nitrogen procedure. *Anal. Chem.* **30**, 1563–1566 (1958).
4. Burtin, P., Doutriaux, D., and Pocidalo, J. J., Contribution a l'étude de la proteinorache. Intérêt de la methode du biuret. *Presse Med.* **66**, 413–416 (1958).
5. Rieder, H. P., Vergleich einiger Methoden zur Bestimmung des Gesamteiweisses im Liquor und anderen stark verdünten Lösungen. *Clin. Chim. Acta* **3**, 455–470 (1958).
6. Burtin, P., Etudes sur les proteines du liquide cephalo-rachidien. *Clin. Chim. Acta* **4**, 72–78 (1959).
7. Lang, C. A., Simple microdetermination of Kjeldahl nitrogen in biological materials. *Anal. Chem.* **30**, 1692–1694 (1958).
8. Daughaday, W. H., Lowry, O. H., Rosebrough, N. J., and Fields, W. S., Determination of cerebrospinal fluid protein with the Folin phenol reagent. *J. Lab. Clin. Med.* **39**, 663–665 (1952).
9. Sevensmark, O., Determination of protein in cerebrospinal fluid: A comment on the Lowry method, *Scand. J. Clin. Lab. Invest.* **10**, 50–52 (1958).
10. Zondag, H. A. and van Boetzelaer, G. L., Determination of protein in cerebrospinal fluid; sources of error in the Lowry method. *Clin. Chim. Acta* **5**, 155–156 (1960).
11. Rieder, H. P., Über Wert und Zuverlässigkeit der Kupfer-Folinmethode zur Bestimmung Kleinster Eiweissmenger. *Clin. Chim. Acta* **6**, 188–194 (1961).
12. Rice, E. W. and Loftis, J. W., Critique of the determination of proteins in cerebrospinal fluid: Evaluation of the biuret method of Goa and the TCA-turbidimetric method of Meulemans. *Clin. Chem.* **8**, 56–61 (1962).
13. Meulemans, O., Determination of total protein in spinal fluid with sulfosalicylic acid and trichloroacetic acid. *Clin. Chim. Acta* **5**, 757–761 (1960).
14. Bossak, H. N., Rosenberg, A. A. and Harris, A., A quantitative turbidimetric method for the determination of spinal fluid protein. *J. Venereal Disease Inform.* **30**, 100–103 (1949).
15. Henry, R. J., Sobel, C. and Segalove, M., Turbidimetric determination of proteins with sulfosalicylic and trichloroacetic acids. *Proc. Soc. Exptl. Biol. Med.* **92**, 748–751 (1956).
16. Cohn, E. J., Hughes, W. L., J. and Weare, J. H., Crystallization of serum albumins from ethanol-water mixtures. *J. Am. Chem. Soc.* **69**, 1753–1761 (1947).
17. Green, J. B., Recent advances in the chemistry of the cerebrospinal fluid. *J. Nervous Mental Disease* **127**, 359–373 (1958).

SALICYLATE*

Submitted by: RODERICK P. MACDONALD, Harper Hospital, Detroit, Michigan
Checked by: JOSEPH H. BOUTWELL JR. and WANDA WILKES, Temple University
               School of Medicine and Hospital, Philadelphia, Pennsylvania
               E. DOYLE SLIFER, Decatur and Macon County Hospital, Decatur,
               Illinois
               ELIZABETH B. SOLOW, Indiana University Medical Center,
               Indianapolis, Indiana
               MILTON STERN, The Mount Sinai Hospital of Cleveland, Cleve-
               land, Ohio

Introduction

A procedure for the determination of salicylate in blood previously appeared in Volume 3 of this series (1). The method, submitted by Routh and Dryer, is based on a modification of the method of Brodie (2), and is an excellent choice for the accurate determination of salicylate. The procedure requires two extraction steps: salicylate from a HCl protein-free filtrate into ethylene chloride, and from ethylene chloride into a ferric nitrate color reagent.

This technique, though accurate at low values, has three disadvantages. (1) It is time consuming, (2) it requires mechanical agitation, and (3) it is an extraction procedure, a process not generally suitable for ultramicro analysis. Since most cases of possible salicylate intoxication occur in young children, an ultramicro procedure is highly desirable. Time is important in these cases, and a shorter, simpler, though admittedly somewhat less accurate method in the low range of serum levels, would be valuable for many laboratories.

Principles

The micro and ultramicro methods presented here are based on the technique proposed by Trinder (3). This utilizes the phenolic nature of salicylate to react with a ferric salt to form a purple colored complex.

* Based on the method of Trinder (3).

237

Salicylic acid

The color inhibition by phosphates and oxalates is eliminated by using a high concentration of ferric nitrate. Mercuric chloride and hydrochloric acid are used to precipitate the serum proteins. The micro modification is by Caraway (4) and the ultramicro modification by Knights, MacDonald, and Ploompuu (5).

Reagents

1. Trinder's reagent. Transfer 40 g. ferric nitrate $[Fe(NO_3)_3 \cdot 9H_2O]$ A.C.S. and 40 g. mercuric chloride ($HgCl_2$) into a 1 l. volumetric flask. Add 120 ml. of 1 N hydrochloric acid, dissolve, and dilute to mark with water. Filter. This reagent is stable indefinitely at room temperature.

2. Salicylic acid stock standard solution, 200 mg. salicylic acid per 100 ml. Transfer 580 mg. of sodium salicylate ($HOC_6H_4CO_2Na$) to a 250 ml. volumetric flask. Dissolve and dilute to mark with water. Add a few drops of chloroform as a preservative. This solution keeps 6 months in a refrigerator at 4–10°C.

NOTE: The Submitter originally proposed the use of salicylic acid as an alternate standard. Checkers (J. H. B. and W. W.) found difficulty in maintaining the required concentration in solution unless the pH was brought to about 7.0 with sodium hydroxide. Therefore, the use of this compound as a standard is not recommended.

3. Salicylic acid, working standard solution, 20 mg. salicylic acid per 100 ml. Dilute 10 ml. of the stock standard to 100 ml. with water. Add a few drops of chloroform as a preservative and keep in a refrigerator. This solution is stable for about 6 months.

NOTE: If only the ultramicro procedure is used, it is more practical to make up only 0.1 of the specified total volume of reagents.

Procedure for Serum or Plasma

MICRO

1. To 0.10 ml. of serum or plasma in a 13 × 100 mm. test tube add 0.90 ml. of water and mix. Prepare a blank and standard solution

using water and working standard in place of serum. These may be prepared directly in cuvets.

2. Add 1.00 ml. of Trinder's reagent, mix well by tapping or by using a vortex mixer, and allow the tubes to stand for 5 minutes.

3. Centrifuge the tubes at high speed for 10 minutes. Decant the clear supernatant into a 12 × 75 mm. cuvet.

NOTE: Checker (E. B. S.) obtained a discrepancy in results for one sample and found that after centrifugation the supernatant solution had a surface scum. She suggests that the supernatant solution be aspirated by capillary pipet rather than by decanting. The Submitter has not had this problem, nor did the other Checkers. It may possibly be explained by a relatively slow speed of centrifugation. The analyst should be alert for this problem and observe the clarity of each supernatant solution.

4. Read the absorbance of the unknown and standard at 540 mμ using the blank solution as a reference. The color developed during the procedure is stable for at least 1 hour (M. S.). If the absorbance reading of the unknown is above the upper limit of linearity, repeat the entire procedure using serum diluted with an appropriate volume of water. Multiply the final concentration result by the dilution factor.

ULTRAMICRO

1. Pipet 250 μl. of water into each of three 1 ml. micro centrifuge tubes.

2. Into one tube add 25 μl. of water (blank), the second tube 25 μl. of serum (unknown), and 25 μl. of working standard into the third tube. Mix well.

3. To each tube add 250 μl. of Trinder's reagent, mix thoroughly, and allow to stand for 5 minutes.

4. Centrifuge at high speed for 5 minutes.

5. Transfer the clear supernatant fluids into cuvets capable of measuring the absorbance of this volume of solution. The color developed during the procedure is stable for at least one hour (M. S.).

NOTE: The Submitter uses 25 × 2.5 × 10 mm. Bessey-Lowry cuvets which require a volume of 0.075 ml.[1]
NOTE: Trinder (3) also used whole blood as a specimen in this procedure.

[1] These cuvets and diaphragm attachment for the Beckman model DU spectrophotometer are obtainable from the Pyrocell Mfg. Co., 91 Carver Ave., Westwood, New Jersey. Such attachments are also available for other instruments.

Editor (S. R. G.) reports that there is an unequal distribution of salicylate between blood cells and plasma water after standing one-half hour.

Procedure for Urine

Dilute urine with water so that it contains 10–40 mg. salicylic acid per 100 ml. Follow the procedure for analysis of serum. After reading the unknown, obtain a urine blank value by setting the instrument at 0 absorbance with water and reading the absorbance of solution prepared by adding 0.10 ml. of diluted urine to 1.00 ml. of Trinder's reagent and 0.10 ml. of phosphoric acid (85–87% H_3PO_4). If centrifugation is necessary, centrifuge both unknown and blank. The addition of phosphoric acid inhibits the chromogenic reaction, presumably by lowering the pH.

Calculations

The Beer-Lambert law may be applied when the absorbance of the unknown is in the range of linearity determined under Calibration.

$$C_{unk} = C_{std} \times \frac{A_{unk}}{A_{std}}$$

where A_{std} is the absorbance of the standard, A_{unk} the absorbance of the unknown, and C_{std} the concentration of the standard in milligrams per 100 ml. (in this case, 20 mg. salicylic acid per 100 ml.). Dilution factors should be applied, when applicable. Values are reported as salicylic acid regardless of the form of salicylate which has been ingested.

For urine:

$$C_{unk} = C_{std} \times \frac{A_{unk_1} - A_{unk_2}}{A_{std}} \times d$$

where C_{unk} is the concentration of salicylic acid in milligrams per 100 ml. urine, C_{std} the concentration of the standard, A_{unk_1} the absorbance of the unknown solution, A_{unk_2} the concentration of the unknown solution after treatment with H_3PO_4, and d the dilution factor.

Calibration

1. Dilute 0.5, 2.5, 5.0, 10.0, 15.0, 20.0, 25.0, and 30.0 ml. of stock standard to 100 ml. with water and add a few drops of chloroform if

the standards are to be stored. These are equivalent to 1, 5, 10, 20, 30, 40, 50, and 60 mg. salicylic acid per 100 ml.

2. Follow the directions given under the procedure for serum and determine the absorbance for each concentration. Plot a standard curve and note the extent of linearity of the curve. A 20 mg./100 ml. standard should be run with each set of unknowns and its absorbance used for calculations. The standard curve will establish the sensitivity and linearity of the method for a particular instrument.

Normal Values

There should be no salicylate present in blood unless the patient has ingested the drug. However, using this method, apparent values of up to 2 mg./100 ml. may be obtained on subjects who have had no salicylate intake.

Discussion

This method meets the requirements set forth in the introduction. It is a simple and rapid method of determining toxic levels of salicylate. Other approaches to salicylate measurement are by differential ultraviolet spectrophotometric analysis (6), single peak ultraviolet absorption (7), use of nitric acid followed by a strong alkali (8), and coupling with diazotized p-nitroaniline (9). None of these methods match the simplicity and rapidity of the Trinder method.

Ultramicro and micro adaptations are presented because salicylate intoxication is primarily a pediatric problem. The Trinder technique is ideally suited for this purpose. Trinder found values of less than 1.1 mg. per 100 ml. of serum and less than 4.5 mg. per 100 ml. of urine from subjects not known to have ingested salicylates. Salicylate values were not affected by addition of 20 mg. bilirubin, 25 mg. phenol, 10,000 I.U. heparin, 1000 mg. glucose, and 1000 mg. urea (all per 100 ml.). Addition of Wintrobe's anticoagulant (150 mg. of ammonium oxalate and 100 mg. potassium oxalate per 100 ml. of serum) gave blank values of 0.3 mg., and 50 mg. ethyl acetoacetate resulted in values of 1.0 mg. per 100 ml.

The ultramicro and micro procedures gave linear calibration curves at least to a concentration of 150 mg. salicylate per 100 ml.

NOTE: The Submitter obtained the following absorbance values for the ultramicro technique (salicylate concentrations in mg./100 ml. are in parentheses): (10) 0.060, (20) 0.105, (30) 0.156, (40) 0.217, (50) 0.255, and (60) 0.317.

The values obtained from the micro technique were: (10) 0.051, (20) 0.105, (30) 0.158, (40) 0.211, (50) 0.260, (60) 0.323, (80) 0.432, and (100) 0.523.

Checkers (J. H. B. and W. W.) reported average recovery of 101% for salicylate added to serum. Checker (M. S.) obtained a low percentage of recovery for a level under 10 mg./100 ml. (5.0 mg/100 ml. added, 3.8 mg./100 recovered), excellent recovery between 10–30 mg./100 ml., and within 2% recovery from 30–60 mg./100 ml., of added salicylate. Similar results were obtained by Checker (E. D. S.).

The proposed technique gives good correlation with the Brodie method (2) (J. H. B. and W. W.) and shows better recovery than the Keller (10) procedure (M. S.).

In an attempt to determine whether the method could be more flexibly described, Checkers (J. H. B. and W. W.) added varying volumes of 1.5 to 7.0 ml. of Trinder's reagent to a constant volume of standard, as well as to serum to which salicylate had been added to obtain a value of 20 mg./100 ml. Results are shown in Table I. Hemolysis and lipemia do not interfere with the procedure (M. S.).

TABLE I

EFFECT OF VARYING VOLUMES OF TRINDER'S REAGENT ON A SERUM
SAMPLE CONTAINING 20 MG./100 ML. SALICYLATE

Serum (ml.)	Water (ml.)	Trinder's reagent (ml.)	Recovered Salicylate (mg./100 ml.)
0.1	0.9	1.5	19.0
0.1	0.9	2.0	20.0
0.1	0.9	3.0	20.3
0.1	0.9	4.0	19.0
0.1	0.9	5.0	18.8
0.1	0.9	6.0	19.6
0.1	0.9	7.0	20.6

The time interval between ingestion of the drug and withdrawal of the blood for analysis is important. It is very often useful to determine the theoretical initial salicylate level. This may be estimated with the equation proposed by Done (11):

$$\log S_I = \log S_s + 0.015T$$

when S_I is the initial level of salicylate, S_s is the serum salicylate in milligrams per 100 ml., and T is the time in hours between ingestion and withdrawal of blood for analysis. Done suggests the following effects for calculated initial salicylate levels: to 50 mg./100 ml., asymptomatic; 50–80, mild intoxication; 80–100, moderate intoxication; 110, severe intoxication, and 160, usually fatal.

SAMPLE CALCULATION

Serum sample obtained 3.5 hours after ingestion of aspirin contained 15 mg. salicylate per 100 ml.

$$\log S_I = \log 15 + 0.015 \times 3.5$$
$$= 1.1761 + 0.0525$$
$$= 1.2286$$
$$S_I = 17 \text{ mg. salicylate per 100 ml.}$$

REFERENCES

1. Routh, J. I., and Dryer, R. L., Salicylate. In "Standard Methods of Clinical Chemistry" (D. Seligson, ed.), Vol. 3, pp. 194–199. Academic Press, New York, 1961.
2. Brodie, B. B., Udenfriend, S., and Coburn, A. F., The determination of salicylic acid in plasma. J. Pharmacol. Exptl. Therap. 80, 114–117 (1944).
3. Trinder, P., Rapid determination of salicylate in biological materials. Biochem. J. 57, 301–303 (1954).
4. Caraway, W. T., "Microchemical Methods for Blood Analysis," pp. 105–106. Thomas, Springfield, Illinois, 1960.
5. Knights, E. M., Jr., MacDonald, R. P., and Ploompuu, J., "Ultramicro Methods for Clinical Laboratories," 2nd ed., pp. 171–175. Grune & Stratton, New York, 1962.
6. Williams, L. A., Linn, R. A., and Zak, B., Differential spectrophotometric determination of serum salicylates. J. Lab. Clin. Med. 53, 156–162 (1959).
7. Stevenson, G. W., Rapid ultraviolet spectrophotometric determination of salicylate in blood. Anal. Chem. 32, 1522–1525 (1960).
8. MacDonald, R. P., Ploompuu, J., and Knights, E. M., Jr., Ultramicro determination of salicylates in blood serum. Pediatrics 20, 515–516 (1957).
9. Moss, D. G., A micro method for blood salicylate estimations. J. Clin. Pathol. 5, 208–211 (1952).
10. Keller, W. J., Jr., A rapid method for the determination of salicylates in serum or plasma. Am. J. Clin. Pathol. 17, 415 (1947).
11. Done, A. K., Salicylate intoxication. Significance of measurements of salicylate in blood in cases of acute ingestion. Pediatrics 26, 800–807 (1960).

UREA NITROGEN AND URINARY AMMONIA

Submitted by: ALEX KAPLAN, Biochemistry Department and University Hospital, University of Washington, Seattle, Washington
Checked by: ALBERT L. CHANEY, Albert L. Chaney Chemical Laboratory, Glendale, California
ROBERT L. LYNCH, Medical College of Virginia, Richmond, Virginia
SAMUEL MEITES, The Children's Hospital, Columbus, Ohio

Introduction

In 1960, Fawcett and Scott (1) introduced a rapid, sensitive and precise method for the determination of blood urea nitrogen by combining the specificity of the urease reaction with a very sensitive colorimetric procedure for the determination of ammonia. Although Berthelot (2) announced in 1859 that ammonia reacts with sodium phenoxide and hypochlorite to produce indophenol blue, more than fifty years elapsed before the reaction was incorporated into a method to quantitate ammonia in biological fluids. In 1913, Thomas (3) measured the ammonia concentration of cerebrospinal fluid by means of hypochlorite-sodium phenoxide reagents. Later workers (4, 5, 6, 7) modified the method and applied it to the estimation of ammonia in body fluids and Kjeldahl digests, as well as to ammonia derived from the action of urease on urea. The final color, however, was not stable, and reproducibility was a problem. Lubochinsky and Zalta (8) introduced the next technical advance with the discovery that sodium nitroprusside catalyzed the Berthelot reaction, improving the intensity, reproducibility, and stability of the blue color. Brown, Duda, Korkes, and Handler (9) modified the concentrations of reagents used by Lubochinsky and Zalta for the determination of ammonia in plasma and tissues, while Fawcett and Scott (1) adapted the color reaction to the determination of serum urea nitrogen. Chaney and Marbach (10) simplified the procedure of Fawcett and Scott by combining the nitroprusside and phenol into one solution, the sodium hydroxide and hypochlorite into a second, thereby reducing the number of reagents from three to two and increasing their stability.

The method presented here is not only very simple and sensitive

245

for manual operation but is readily adaptable to automated procedures.[1] Furthermore, the color reaction may be utilized for the
direct determination of submicrogram quantities of ammonia in body
fluids or solutions.

Principle

Urea nitrogen is determined in two steps:
1. Hydrolysis of urea to ammonia and carbon dioxide by the
enzyme, urease.

$$\begin{array}{c} H_2N \\ \diagdown \\ C{=}O + H_2O \xrightarrow{\text{urease}} 2NH_3 + CO_2 \\ \diagup \\ H_2N \end{array}$$

2. Determination of ammonia nitrogen by the Berthelot reaction.
When sodium nitroprusside is used as a catalyst, ammonia reacts
with an alkaline solution of sodium hypochlorite and phenol to form
an intensely blue indophenol. The concentration of ammonia is directly proportional to the absorbance of the indophenol which is
measured spectrophotometrically.

(a) $\qquad\qquad\qquad\qquad NH_4^+ + OH^- \leftrightarrow NH_3 + H_2O$

(b) 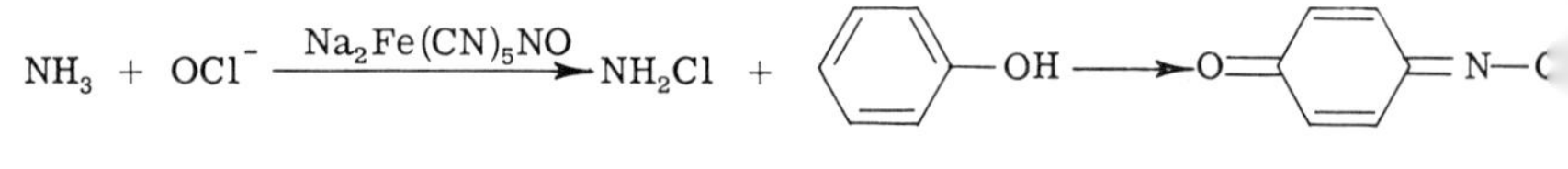 $NH_3 + OCl^- \xrightarrow{Na_2Fe(CN)_5NO} NH_2Cl$ +

                    Chloroamine   Phenol   Quinonechloroamine

(c)

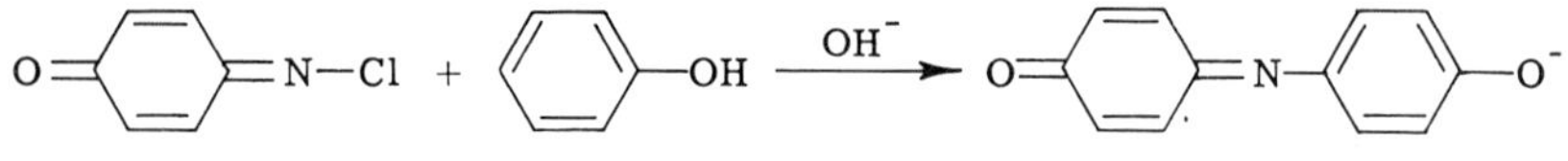

                                 Indophenol
                        (dissociated form, blue dye)

The net reaction is:

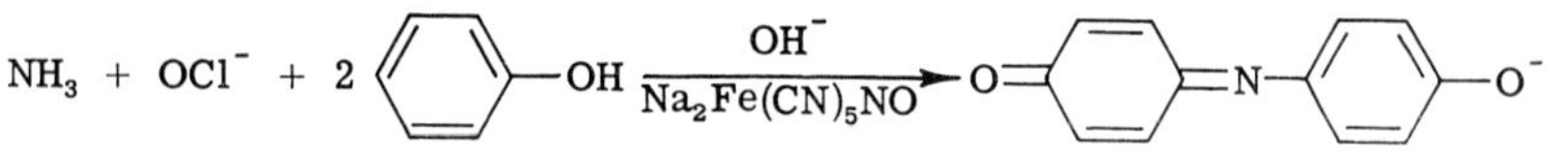

---

[1] The procedure has been adapted to the following automated systems: (a)
Autoanalyzer, Technicon Instruments Corporation, Chauncey, New York, by Wilcox and Sterling (11). (b) Robot Chemist, Research Specialties Co., Richmond, California, Procedure 14A; modification by Kaplan and Snook (12).

The exact step at which sodium nitroprusside acts as a catalyst is not known with certainty but is probably involved in the formation of quinonechloroamine shown in reaction (b). Bolleter, Bushman, and Tidwell (13) have reported that p-aminophenol reacts with phenol in alkaline solution in the absence of hypochlorite or a catalyst to produce a blue compound with the same light absorption characteristics as that of indophenol derived from ammonia. This has been confirmed in the Submitter's laboratory (12); the blue dye forms immediately at room temperature upon the addition of p-aminophenol to an alkaline phenol solution.

NOTE: The color of the indophenol dye is pH dependent (14). In acid solution, indophenol exists in the undissociated form, which is yellow, while in alkaline solution it dissociates as shown in reaction (c) with a change to a blue color.

NOTE: For manual methods, a slight modification of the Chaney and Marbach (10) procedure, using the dilute reagents, is recommended. The method as outlined has been employed in the Submitter's laboratory for more than a year, while prior to the simplification of reagents, the method proposed by Fawcett and Scott (1) was used for 2 years.

Reagents

NOTE: Distilled water for the preparation of reagents and blanks should be rendered free of ammonia by passing it through an ion-exchange column. High blanks may be obtained if this is not done.

1. Stock phenol-nitroprusside solution. To approximately 500 ml. of distilled water in a liter volumetric flask, add 50 g. of reagent grade phenol and 0.25 g. of sodium nitroprusside,[2] $Na_2Fe(CN)_5NO \cdot 2H_2O$. Dilute to the mark with distilled water and transfer to an amber bottle. The reagent is stable for at least 2 months when stored at 4–10°C. The solution contains 0.5 M phenol and 0.001 M sodium nitroprusside.

NOTE: Deviations of ±20% in concentrations of phenol and nitroprusside do not affect the color intensity of the final reaction mixture. When the concentration of phenol is reduced by 40%, however, the intensity of color produced in the reaction by a given amount of ammonia is reduced and the reaction no longer obeys Beer's law (12).

NOTE: Use reagent grade phenol. Discard the phenol solution if it turns brown.

[2] Listed by most American chemical companies as sodium *nitroferricyanide*. Both names are incorrect from a chemical standpoint since the product is a *nitroso-* and not a *nitro* compound. According to modern nomenclature (15), the proper name is sodium pentacyanonitrosyloferrate (III).

2. *Dilute phenol-nitroprusside solution.*[3] Dilute 250 ml. of the stock solution to 1 l. with distilled water. Store the solution in an amber bottle at 4–10°C.

NOTE: This is a slight departure from the procedure of Chaney and Marbach (10). They make a 1:5 dilution and use 5 ml. whereas the Submitter uses 4 ml. of 1:4 dilution, resulting in the same concentration of reagents in the reaction mixtures. This modification is introduced to change the final volume of the analysis to 10 ml.

3. *Stock alkaline hypochlorite solution.* In approximately 600 ml. of distilled water in a 1 l. volumetric flask, place 25 g. of NaOH pellets. When cool, add 40 ml. of hypochlorite solution[4] containing 5% by weight of NaOCl. Dilute to volume with water. The reagent is stable for at least 3 months when it is protected from light and stored at 4–10°C.

NOTE: Neither the NaOH nor NaOCl concentrations are critical. Reagents were tested containing varying amounts of hypochlorite. When standard solutions containing 1.5 to 20 μg. of NH_3-N were run, same results were obtained when the volume of hypochlorite solution for the stock solution was varied from 20 ml. to 55 ml. (12).

4. *Dilute alkaline hypochlorite solution.*[3] Dilute 200 ml. of stock solution to 1 l. with distilled water. Store in an amber bottle at 4–10°C. where it is stable for at least 1 month.

5. *EDTA, 1% (w/v), pH 6.5.* Dissolve 20 g. of the disodium salt of ethylenediaminetetraacetic acid in approximately 1.5 l. of distilled water. Adjust the pH to 6.5 by the addition of 1 N NaOH. Make up to 2 l. with water.

NOTE: Urease is readily inactivated by traces of heavy metals. EDTA forms complexes with ions of heavy metals and provides insurance against this type of inactivation. When adjusted to pH 6.5, the EDTA solution also serves as a buffer. The incubation medium should be slightly acid in order to avoid possible loss of NH_3.

6. *Stock urease.* Dissolve 500 mg. of special purity urease[5] (activity 3500–4100 U./g.) in 25 ml. of distilled water and add 25 ml. of glycerol. The solution is stable for at least 4 months at 4–10°C.

[3] Concentrated solutions (UN-Test) are available as a kit from Hyland Laboratories, Los Angeles 39, California, which, when diluted to 500 ml. volume, contain the same reagent concentrations as the dilute solutions of Chaney and Marbach (10).

[4] Three commercial brands of sodium hypochlorite bleach (Clorox, White, and West Best) were found to be satisfactory. Other brands were not tested, but many could probably be used.

[5] Type V urease. Obtained from Sigma Chemical Co., St. Louis 18, Missouri.

NOTE: One unit (modified Sumner) is defined as that amount of enzyme which will liberate 1 mg. of ammonia nitrogen from urea in 5 minutes at pH 7.0 and 30°C.

NOTE: Two grams of urease[6] containing 800–1000 U./g. may be substituted for the special purity urease, with approximately a doubling of the absorbance of the blank. The cost per 1000 units of enzyme activity is one-fourth that of the special purity type.

7. *Dilute urease solution, 0.4 U./ml.* Dilute 1 ml. of stock urease to 100 ml. with EDTA solution. The solution is stable for at least 3 weeks when stored at 4–10°C.

NOTE: Pour into a small flask the amount of enzyme solution sufficient for the day's work. Discard the excess at the end of the day.

8. *Stock urea nitrogen standard, 500 mg./100 ml.* Dry reagent grade urea in a vacuum desiccator. Dissolve 1.0717 g. urea in distilled water in a 100 ml. volumetric flask. Add 0.10 g. of sodium azide as a preservative. Dilute to volume with water and store at 4–10°C.

NOTE: Sodium azide prevents the growth of microorganisms but does not inhibit urease action nor interfere with the color reaction; standard solutions are stabilized for at least 6 months.

9. *Working urea nitrogen standards:*
(*a*) *15 mg./100 ml.* Dilute 3.00 ml. of stock urea-N standard to 100 ml. with 0.1% (w/v) sodium azide solution. Store at 4–10°C.
(*b*) *50 mg./100 ml.* Dilute 10.0 ml. of stock urea-N standard to 100 ml. with 0.1% (w/v) sodium azide solution. Store at 4–10°C.
10. *NH₃-N stock standard, 500 mg./100 ml.* Dissolve 2.3581 g. of desiccated $(NH_4)_2SO_4$ in distilled water and make up to 100 ml.
11. *Working NH₃-N standard, 50 mg./100 ml.* Dilute 10.0 ml. of stock NH₃-N standard to 100 ml. with distilled water. Store at 4–10°C.

Procedures

UREA NITROGEN IN SERUM

NOTE. Serum is recommended for this test as it is the most convenient fluid to obtain for routine use, but plasma may be substituted provided the anticoagulant contains no ammonium salts. Sodium salts of heparin, ethylenediaminetetraacetate, or oxalate are satisfactory. It is not necessary to precipitate serum

[6] Type II urease. Obtained from Sigma Chemical Co., St. Louis 18, Missouri. Urease in Hyland UN Test kit is satisfactory.

or plasma proteins. Whole blood cannot be used since the high concentration of hemoglobin would seriously interfere by forming a brown precipitate.

NOTE: The procedure should be carried out in a room free of traces of ammonia vapors. It is inadvisable to store urine in the same room. All operations involving the use of NH_4OH or its generation by the addition of alkali to ammonium salts (e.g., in the sodium nitroprusside test for ketones), should be conducted in a fume hood or in another room.

1. Pipet the following volumes into 16 × 100 or 150 mm. test tubes:

	Unknown (ml.)	Standard (ml.)	Reagent blank (ml.)
Urease	1.00	1.00	1.00
Serum	0.010	—	—
Working Standard	—	0.010	—

NOTE: It is very convenient to pipet the samples (Unknown, Standards, Control) and transfer them to the appropriate test tube by wash-out with 0.8 ml. of water by means of commercially available pipetter-dispensers.[7] If such a pipetter is used, the dilute urease solution is made 5 times stronger (5 ml. of stock urease diluted to 100 ml. with EDTA solution) and 0.2 ml. is taken instead of 1 ml. as above.

NOTE: For rigid control, it is advisable to include in each run urea-N standards (15 and 50 mg./100 ml.), NH_3-N standard (50 mg. per 100 ml.), and a control serum which is treated as an unknown.

2. Incubate the mixtures for 15 minutes at 37°C. or 5 minutes at 50–55°C.

3. Add, in rapid succession, 4.0 ml. of dilute phenol-nitroprusside solution and 5.0 ml. of dilute alkaline hypochlorite solution to each tube.

NOTE: Both solutions must be added to each tube in the order named and mixed before proceeding to the next one.

4. Place the tubes in 37°C. water bath for 20 minutes to develop color.

5. Transfer the contents of each tube to a 10 or 12 mm. cuvet and read the absorbance at 560 mμ, using the blank as reference.[8]

[7] Dilumat, produced by Research Specialties Co., Richmond, California, or Auto-Dilutor of Scientific Products.

[8] The blue color is so intense that it is not practical to use the optimum wavelength of 628 mμ because the range of measurable urea-N would be restricted. The absorbance at 560 mμ is approximately 50% of that at 628 mμ. Using the Coleman Jr. spectrophotometer, model 6A, 12 mm. cuvets and a wavelength of 560 mμ, the net absorbance produced by 5 μg. of NH_3-N (equivalent to 10 μl.

NOTE: Always check the absorbance of the blank against water to see whether any of the reagents have absorbed NH_3. The absorbance of the blank referred to water should not exceed 0.040 using special purity urease or 0.100 for the less potent preparation.

NOTE: Including an NH_3-N standard with the urea standards aids in locating the source of trouble if the blue color should fail to develop properly. Full color in the NH_3-N standard but decreased color in the urea standards and unknown indicates that the urease is inactive. Decreased color in all tubes indicates that one or more reagents is deteriorated. This happens rarely but the alkaline hypochlorite solution is the first one to suspect. Prolonged exposure to light is the most common cause of deterioration.

NOTE: The alternate procedure outlined by Chaney and Marbach (10) of adding 1 ml. each of the respective concentrated solutions instead of dilute reagents works just as well. The color reaction is completed in approximately 3 minutes at 60°C. and 5 minutes at 37°C., giving the same absorbance per microgram of NH_3-N as does the dilute reagents. Though the reaction time for color development is shorter with the concentrated reagents, the procedure with the dilute reagents is preferred for routine manual operation because one less step is required. The addition of 7 ml. of water after color development is required when using the concentrated reagents in order to adjust the final volume to 10 ml. It is advantageous to employ the concentrated reagents in emergencies and in automated methods (12) because the test is finished sooner.

6. Calculation. The absorbance values for the two aqueous urea standards should be consistent and are employed in the calculation,

$$\text{mg. urea-N}/100 \text{ ml. of serum} = \frac{A_{unk}}{A_{std}} \times \text{conc. of standard}$$

where A is absorbance.

URINE UREA-N AND PREFORMED AMMONIA-N

As ammonium salts are usually present in urine, the use of a urease method for the determination of urine urea-N requires either the prior removal of preformed ammonia or its determination and the subtraction of NH_3-N from the concentration of urea-N $+ NH_3$-N. The latter approach is simpler to carry out.

1. For the determination of urea-N $+ NH_3$-N, dilute the urine 1:10 with water. Use undiluted urine for the measurement of NH_3-N.

2. Transfer to 16 × 150 mm. test tubes the following:

---

of serum containing 50 mg./100 ml. of urea-N) is approximately 0.350. Thus, concentrations of urea-N from 5 to 100 mg./100 ml. can be read directly. Higher values can be read by diluting the blank and unknown with up to 3 vol. of water, extending the upper range of urea-N to 300 mg./100 ml. without the necessity of repeating the test.

	Urea-N			NH$_3$-N		
	Tube A for urea-N + NH$_3$-N	Tube B for urea-N std.	Tube C for urea-N blank	Tube D for NH$_3$-N	Tube E for NH$_3$-N std.	Tube F for NH$_3$-N blank
Urease (ml.)	1.00	1.00	1.00	—	—	—
Water (ml.)	—	—	—	1.00	1.00	1.00
Urine (ml.)	0.010	—	—	0.010	—	—
Urea-N wkg. std. (ml.)	—	0.010	—	—	—	—
NH$_3$-N wkg. std. (ml.)	—	—	—	—	0.010	—

3. For Tubes A, B, and C, carry out steps 2 through 6 described for serum urea-N. Calculate the concentration of urea-N + NH$_3$-N in milligrams per 100 ml., taking into consideration the dilution of the urine.

4. For Tubes D, E, and F, carry out steps 3 through 6. Calculate the concentration of NH$_3$-N in milligrams per 100 ml.

5. The concentration of urea-N is obtained by subtracting the concentration of NH$_3$-N from that of urea-N + NH$_3$-N.

Standardization

UREA-N

1. Prepare a series of urea-N standards by diluting the stock urea-N standard, 500 mg./100 ml., with water as described below, using volumetric flasks.

Flask No.	Volume flask (ml.)	Volume std. urea-N (ml.)	Conc. urea-N (mg./100 ml.)
1	50 (blank)	0.0	0.0
2	50.0	1.00	10.0
3	10.0	0.50	25.0
4	10.0	1.00	50.0
5	10.0	2.00	100.0

2. Carry each standard through the procedure for serum urea-N, steps 1 through 6, using 0.010 ml. of each standard.

3. Plot the absorbance-concentration curve on rectangular graph paper.

NH_3-N

Substitute the NH_3-N stock standard, 500 mg./100 ml., for the urea-N standard and make the same series of dilutions as described above. Substitute water for the urease solution and omit the incubation in step 2 but follow steps 3 through 6. A calibration curve similar to that of urea-N is obtained.

Discussion

The application of the catalyzed Berthelot reaction with ammonia to the determination of urea-N has many advantages over other methods. The method is extremely sensitive, and as outlined, readily measures the urea-N in 10 μl. of normal serum (equivalent to 0.7–1.8 μg. of N). The amount of N readily quantitated could be reduced to 0.1 μg. by reducing the final volume of solution to 5 ml. or less and measuring the light absorbance at 628 mμ, the optimum wavelength. The only ultramicro equipment necessary is the pipet for the measurement of the serum sample. A wide range of concentrations is covered by the method. The reaction obeys Beer's law from 0.5 to 30 μg. of urea-N; unknown and blanks may be diluted with an equal volume of water when the concentration exceeds this range. The method is rapid, requiring about 40 minutes to complete when using the dilute reagents or 25 minutes with the concentrated solutions. The color is stable for at least 20 hours.

There are few substances normally present in serum or urine that interfere with the measurement of urea-N or NH_3-N by the Chaney and Marbach procedure (10). The following substances have negligible interference: glucosamine, citrulline, bilirubin, glutamine, hemoglobin [Fawcett and Scott (1)]; anthranilic acid, histidine, uracil, glutamine, guanidine, arginine, lysine [Fenton, (16)]; sulfadiazine, alanine, phenylalanine, uric acid, creatinine, salicylic acid [Meites (17)]; aliphatic amines, sodium chloride, potassium nitrate, sodium sulfate, barium chloride, copper, zinc, and iron salts, p-hydroquinone and related salts [Bolleter et al. (13)].

NOTE: Some of these nitrogenous substances, however, have been found to produce the indophenol blue color when the Berthelot reaction is carried out with *different reagent preparations and under different reaction conditions.* Thus, Wearne (18) employed a procedure in which pure solutions or plasma filtrates reacted with chlorine water or hypochlorite at various pH's before adding phenol and alkali. Under these conditions, urea, uric acid, amino acids, and other nitrogenous compounds produced a blue color when the pH of the test

254   ALEX KAPLAN

solution was below 5. When the pH was greater than 7, ammonia was the only substance tested that was chromogenic. Using a different procedure, Bolleter *et al.* (13) boiled their test solutions with chlorine water and phenol and then added alkali after cooling the solutions. Under these conditions, two substances interfered with the reaction: bromide ions caused a precipitate to form (brominated phenol) while hydroxylamine inhibited the Berthelot reaction completely. Kaplan and Snook (12) tested these latter two substances for interference in the Chaney and Marbach procedure (10). Bromide added to serum to a concentration of 500 mg./100 ml. neither caused a precipitate to form nor interfered with the measurement of urea-N. Likewise, the addition of hydroxylamine to serum to a concentration of 500 mg./100 ml. did not inhibit the Berthelot reaction when urea-N was measured.

The only extraneous compounds so far known to produce colors in the Berthelot reaction and cause falsely high values in the determination of urea-N or NH$_3$-N are *p*-aminophenol and substituted *p*-aminophenols (13). These substances are not commonly used in the therapeutics, so their occurrence in body fluids are quite rare. If these compounds are suspected, they can be corrected for by determining a blank in which the urease is added *after* the addition of phenol reagent.

A grossly lipemic serum introduces a turbidity that persists through the color development; absorbance in a 12 mm. cuvet may be increased by 0.010 by this type of light dispersion. The effect of lipemia may be eliminated by extracting the final colored solution with 1 or 2 ml. of ether or compensated for by reading against a serum blank in which the urease is added after the addition of phenol reagent. The ammonia level of plasma is so low that even in hepatic coma it would cause no appreciable error in the determination of serum urea-N. A plasma NH$_3$-N concentration of 5 μg./ml. would raise the apparent urea-N level by 0.5 mg./100 ml. Contamination of the atmosphere with ammonia fumes, however, may introduce a very serious error.

NOTE: A number of laboratories (9, 12, 19, 20, 21) have reported the use of the Berthelot reaction in conjunction with a variety of separation techniques to measure the ammonia concentration of plasma.

Normal and Abnormal Values for Serum Urea-N

The normal range for serum urea-N is the same when measured by the modified Berthelot reaction as that obtained by other reliable methods, a finding that has been documented by Searcy, Gough, Korotzer, and Berquist (22). The normal range of urea-N concentration which embraces 95% of the values, varies from 7 to 18 mg./100 ml. of serum. Its concentration in the water of plasma and erythrocytes is the same, but since the red cells have a lower water content per unit volume, the concentration of urea-N in whole blood is about 15% lower than that of plasma or serum. Within limits, the serum concentration of urea-N varies with the protein content of the diet; it can be doubled by the ingestion of diets very high in protein.

A rise in the level of serum urea-N may be caused by an impairment in the excretion of urea, by an accelerated production as a result of increased protein catabolism, or by a combination of both factors. Since the formation and excretion of urine depends upon an adequate blood flow and pressure in the kidney, a sufficient number of normally functioning renal nephrons, an adequate collecting system, and unobstructed excretory channels, many different pathological conditions may lead to impaired urine formation or excretion. Thus, an elevated serum urea-N may result from causes that are pre-renal, renal, or post-renal in origin. The discussion of abnormal results is treated more extensively elsewhere (23).

Urine Urea-N and NH_3-N

Urea is the nitrogenous constituent that is present in the highest concentration in urine. Approximately 15 g. of urea-N is excreted per day but this amount will vary with the dietary content of protein and the functional adequacy of the kidneys.

The determination of urine urea-N is necessary for the measurement of the urea clearance. Aside from this test of renal function, there are few, if any, indications for the measurement of urine urea-N.

Ammonia is also a normal constituent of urine, with approximately 0.5 or 0.6 g. of NH_3-N excreted per day on an average diet. The amount may be greatly increased during conditions of acidosis or following the ingestion of acid-forming foods.

As mentioned earlier, it is necessary to either measure the NH_3-N in urine or adsorb it (with Permutit) when measuring the concentration of urea-N in urine. It is seldom necessary to determine only the NH_3-N content of urine, although it may be requested in conjunction with the measurement of titratable acidity. The sum of ammonia excretion and titratable acidity is an indicator of the renal capacity to conserve base.

REFERENCES

1. Fawcett, J. K., and Scott, J. E., A rapid and precise method for the determination of urea. *J. Clin. Pathol.* 13, 156–159 (1960).
2. Berthelot, M., Répertoire de Chimie appliqué. 1, 284 (1859).

3. Thomas, P., Recherche et dosage de l'ammoniaque dans le liquide cephalo-rachidien. *Bull. Soc. Chim. France* **13**, 398–400 (1913).
4. Orr, A. P., A colorimetric method for the direct estimation of ammonia in urine. *Biochem. J.* **18**, 806–808 (1924).
5. Van Slyke, D. D., and Hiller, A., Determination of ammonia in blood. *J. Biol. Chem.* **102**, 499–504 (1933).
6. Borsook, H., Micromethods for determination of ammonia, urea, total nitrogen, uric acid, creatinine (and creatine) and allantoin. *J. Biol. Chem.* **110**, 481–493 (1935).
7. Russell, J. A., The colorimetric estimation of small amounts of ammonia by the phenol-hypochlorite reaction. *J. Biol. Chem.* **156**, 457–461 (1944).
8. Lubochinsky, B., and Zalta, J. P., Microdosage colorimétrique de l'azote ammoniacal. *Bull. Soc. Chim. Biol.* **36**, 1363–1366 (1954).
9. Brown, R. H., Duda, G. D., Korkes, S., and Handler, P., A colorimetric micromethod for the determination of ammonia; the ammonia content of rat tissues and human plasma. *Arch. Biochem. Biophys.* **66**, 301–309 (1957).
10. Chaney, A. L., and Marbach, E. P., Modified reagents for determination of urea and ammonia. *Clin. Chem.* **8**, 130–132 (1962).
11. Wilcox, A. A., and Sterling, R. E., The use of the Berthelot reaction in the automated analysis of serum urea. *Clin. Chem.* **8**, 427 (1962).
12. Kaplan, A., and Snook, M., Unpublished observations (1963).
13. Bolleter, W. T., Bushman, C. J., and Tidwell, P. W., Spectrophotometric determination of ammonia as indophenol. *Anal. Chem.* **33**, 592–594 (1961).
14. Rodd, E. H., "Chemistry of Carbon Compounds," vol. III, Part B, pp. 721–722, Elsevier, Amsterdam, 1956.
15. Fernelius, W. C., Nomenclature of coordination compounds and its relation to general inorganic nomenclature. Advances in Chemistry series. American Chemical Society, Washington, D.C., **8**, 9–37 (1953).
16. Fenton, J. C. B., The estimation of plasma ammonia by ion exchange. *Clin. Chim. Acta* **7**, 163–175 (1962).
17. Meites, S., Unpublished observations (1963).
18. Wearne, J. T., Non-specificity of hypochlorite-phenol estimation of ammonium in biological material. *Anal. Chem.* **35**, 327–329 (1963).
19. Ternberg, J. L., and Hershey, F. B., Colorimetric determination of blood ammonia. *J. Lab. Clin. Med.* **56**, 766–776 (1960).
20. Miller, G. E., and Rice, J. D., Jr., Determination of the concentration of ammonia nitrogen in plasma by means of a simple ion exchange method. *Am. J. Clin. Pathol.* **39**, 97–103 (1963).
21. Forman, D. T., Rapid determination of plasma ammonia by an ion exchange technique. *Clin. Chem.* **10**, 497–508 (1964).
22. Searcy, R. L., Gough, G. S., Korotzer, J. L., and Berquist, L. M., Evaluation of a new technique for estimation of urea nitrogen in serum. *Am. J. Med. Technol.* **27**, 255–262 (1961).
23. Cohn, C., and Kaplan, A., *In* "Textbook of Clinical Pathology" (S. E. Miller, ed.), Chapter 7, pp. 242–245. Williams & Wilkins, Baltimore, Maryland, 1960.

XYLOSE*

Submitted by: MIRIAM REINER and HELEN L. CHEUNG, District of Columbia
General Hospital, Washington, District of Columbia
Checked by: SISTER MARY CLAIRE KENNEDY, S.S.J. and OLGA EKIMOFF, St.
Vincent Hospital, Erie, Pennsylvania
SYLVAN M. SAX, Western Pennsylvania Hospital, Pittsburgh,
Pennsylvania
RAYMOND E. VANDERLINDE and JEWELL McCREARY, Memorial
Hospital, Cumberland, Maryland

Introduction

Recent studies of the oral administration of xylose as a measure
of absorption in the small intestine have led to a renewed interest in
the determination of pentoses. Earlier workers measured the total
reducing substances and then, after yeast fermentation, determined
the residual nonglucose reducing constituents after correcting for
nonxylose reducing substances (2). Roberts, Beck, Kallos, and Kahn
(3) have used the same principle replacing the yeast with glucose
oxidase. The method of Roe and Rice (1) for pentoses has been generally adopted for its reliability, simplicity, and speed. They have
improved the specificity for pentoses by replacing strong mineral acids
with acetic acid and employing milder reaction conditions than had
been utilized in other reactions involving furfural as an intermediate.
However, this is achieved with some loss in sensitivity (4). Other
workers (5, 6) have adapted this procedure to micro levels. Crowley
(7) has automated the procedure for xylose.

Other pentose methods are based on their reaction with orcinol
(8, 9), cystine-sulfuric acid (10), carbazole (11), and tetrazolium
(12).

The administration of a large dose of xylose was orginally proposed
as a test of renal function (13). Later, xylose was used as a measure
of absorption from the small intestine by Helmer and Fouts (14)
in pernicious anemia, and Gardner and Perez-Santiago (15) in an
extensive study of sprue. Fourman (16), after observing the reduction
in xylose excretion in patients with sprue, suggested that it could
be used as a measure of absorption in the small intestine and in the

° Based on the method of Roe and Rice (1).

257

diagnosis of related malabsorption syndromes. Subequently, after the studies of Brien, Turner, Watson, and Geddes (17), and Benson, Culver, Ragland, Jones, Drummey, and Bougas (18), there has been an extensive interest in xylose absorption. In many hospitals, the xylose excretion test is considered a routine procedure for the diagnosis of malabsorption syndromes.

Principle

The method of Roe and Rice (1) for determining pentoses in animal tissue has been adapted to blood and urine analyses. The procedure is based on the reaction of p-bromaniline acetate with furfural, formed from the pentose at 70° C., in glacial acetic acid saturated with thiourea. The pink-colored complex formed is measured spectrophotometrically at 520 mμ. The addition of an antioxidant, thiourea, to the reagent and the maintenance of the temperature at 70° C. achieves a high degree of specificity, and minimizes the effect of interfering substances. Other aldopentoses, e.g., arabinose, or ribose, if present, will react similarly; but as the normal excretion of these pentoses is very small, about 6 mg./100 ml. (17), there is usually no interference with the test (19).

The proteins of blood plasma or serum are precipitated according to the method of Somogyi (20) prior to the determination of xylose. According to Park, Reinwein, Henderson, Cadenes, and Morgan (21) and Helmreich and Cori (22), the passage of xylose into the red cell is fairly rapid. Simultaneous determinations on serum and whole blood (Sr. M. C. K.) give similar results, so the choice of specimen is immaterial.

Reagents

1. Glacial acetic acid saturated with thiourea. Prepare a solution of glacial acetic acid (reagent grade) saturated with thiourea (NH_2CSNH_2). Approximately 4 g. thiourea will saturate 100 ml. of glacial acetic acid. Solution is facilitated by frequent agitation either mechanically or by a magnetic stirrer for at least 1 hour. If a stirrer is not available, the solubility of this reagent can be increased by storing it in a flask in a water bath at 37° C. The presence of excess thiourea during storage is desirable. This solution is stable for months.

2. *p-Bromoaniline reagent, 2% (w/v)*. Add 2.0 g. of pure *p*-bromo-aniline ($BrC_6H_4NH_2$),[1] to 100 ml. of reagent (1), glacial acetic acid saturated with thiourea. This reagent is stable for at least 2 weeks if it is refrigerated (4–10°C). Whether it is stored in a clear or dark bottle makes no difference. Many chemists prefer to prepare this reagent just before use.

NOTE: Checker (Sr. M. C. K.) found that if a visible pink color develops this will fade in 2–3 hours. This reagent should be prepared in advance to minimize the reagent blank, which is then constant. (R. E. V.) prefers freshly prepared *p*-bromoaniline. In a comparison with a 2 week old reagent he found that the older reagent intensifies color development very extensively but gives less reproducible results.

3. *Somogyi deproteinizing reagents*.[2]

(*a*) *5% (w/v) zinc sulfate* ($ZnSO_4·7H_2O$). Dissolve 50.0 g. of $ZnSO_4·7H_2O$ in water and dilute to 1 l. in a graduated cylinder. Filter if cloudy.

(*b*) *0.3 N barium hydroxide*. Dissolve 45.0 g. of $Ba(OH)_2·8H_2O$ in water and dilute to 1 l. in a graduated cylinder. Filter if cloudy. Store in well-stoppered containers filled to capacity and protected with soda-lime tubes. Avoid unnecessary exposure to the air and restandardize weekly.

NOTE: Sodium hydroxide, 0.3 *N*, may be preferred to $Ba(OH)_2$ for the Somogyi protein free filtrate. Dissolve 12.0 g. NaOH pellets in water and dilute to 1 l. Filter if cloudy.

NOTE: Absolute concentration of the working solutions of zinc sulfate and barium or sodium hydroxide is not as important as their precise neutralization. To titrate, measure 10.00 ml. of $ZnSO_4$ solution into a 250 ml. flask, add approximately 50 ml. of water and 4 drops of phenolphthalein indicator. *Slowly* titrate with the hydroxide using constant agitation until one drop turns the solution to a faint pink. This should require 10.0 ml. ± 0.05 of hydroxide solution. If necessary, add an appropriate amount of distilled water to the stronger solution, and repeat the titration.

After the addition of the above solutions to blood, plasma, or serum, the filtration should proceed rapidly to give a clear filtrate with little tendency to foam.

---

[1] This may be obtained from Distillation Products Industries.

[2] Checker (Sr. M. C. K.) has found that storing the reagents in collapsible containers eliminates the problem of the formation of precipitates when they come in contact with the atmosphere. Harleco-Handiboy or Fisher Polypak may be used.

4. Xylose standards.

(*a*) *Stock standard.* Dissolve 1.000 g. D(+)-xylose[3] in about 50 ml. of saturated benzoic acid (about 0.25% w/v in water) and make up to volume with the benzoic acid solution in a 100 ml. volumetric flask. This solution is stable indefinitely.

(*b*) *Working xylose standards.*

20 mg. in 100 ml. (0.2 mg./ml.). Dilute the stock standard 1:50 with saturated benzoic acid solution.

10 mg. in 100 ml. (0.1 mg./ml.). Dilute the stock standard 1:100 with saturated benzoic acid solution.

5 mg. in 100 ml. (0.05 mg./ml.). Dilute the stock standard 1:200 with saturated benzoic acid solution.

Procedure

BLOOD

Prepare a protein-free filtrate according to the method of Somogyi (20). Mix 1.00 ml. of blood, plasma or serum, 5.00 ml. of distilled water and 2.00 ml. of hydroxide. Add 2.00 ml. of 5% $ZnSO_4$, mix, and centrifuge at 3000 rpm for 5 min. The filtrate should be prepared without delay after blood collection; otherwise, keep the specimens in the refrigerator.

URINE

NOTE: Ordinarily, preparation of Somogyi filtrates on urine is rarely necessary (23). If the urine is bloody or turbid after centrifugation, or if it contains bilirubin, a 1:10 Somogyi filtrate should be prepared from a 1:10 aqueous dilution.

Checker (S. M. S.) determined xylose on 1:100 dilutions of pentose-free urine specimens including those that were: dark amber, strongly acetone-positive, 3 plus for protein, turbid, bloody, bilirubin containing, post-Diagnex, and post-Pyridium. Only in the cases mentioned above was it found advantageous to prepare Somogyi filtrates. Even then, the elevation resulting from "blank" color was moderate, at most. Significant color was produced only in the post-Pyridium specimen, about the same amount in either aqueous dilution or Somogyi filtrate.

Prepare 1:50 and 1:100 aqueous dilutions of urine. Urine specimens containing xylose are stable at room temperature for at least 48 hours (24) and indefinitely, if refrigerated (18). Kerstell (23) reported

[3] Pure D(+)-xylose from either Fisher Scientific Co. or Pfanstiehl Chemical Co. is satisfactory. Use D(+)-xylose *only*. Never use L(−)-xylose as its rate of absorption is much slower.

it unnecessary to run individual blanks of urine; the Submitters found this true except for the conditions stated in the Note above.

1. Pipet 1.00 ml. of water, standards, blood filtrate, and urine dilutions in a series of duplicate tubes or cuvets. Add 5.00 ml. of p-bromoaniline reagent to *all* tubes. If a distinct pink color develops when the latter is added to the urine samples prior to heating, it usually indicates that a greater dilution should be made.

2. Incubate one of the duplicate tubes in a water bath at 70°C. ± 2° for 10 minutes keeping the water level higher than the liquid level inside the tubes. Cool to room temperature under running water, or a cold water bath. Use the unheated set of tubes as blanks.

NOTE: Checker (S. M. S.) found that day-to-day variation in readings of the standards is reduced by the use of uniform, pyrex, wide test tubes (150 × 20 mm.), which are shaken intermittently in the first few minutes of the short heating cycle.

Checker (R. E. V.) placed xylose standards (0.05 mg./ml. to 0.3 mg./ml.) and blanks in an ice-bath for 10 minutes while the duplicates were heated to 70°C. and found that they consistently gave blank readings less than 0.01 A and usually less than 0.004 A.

3. Place *all* tubes in the dark at room temperature for 70 minutes. Obtain the absorbance reading of each tube in a spectrophotometer or photoelectric colorimeter at 520mμ. The unheated tube serves as a blank for each corresponding heated tube. Read within 30 minutes as the color fades after this time (1). If the reading of the unknown is too high, dilute with the reagent blank and make appropriate calculations.

NOTE: The day to day variation is too great to allow the use of a pre-calibrated standard curve for xylose. If desired, a *daily* curve of working standards containing 5, 10, and 20 mg./100 ml. (corresponding to 50, 100, and 200 mg. when 1:10 dilutions of sample are used) may be constructed and the results read directly; otherwise, compare the absorbance of the unknown with that of the nearest standard.

The choice of the range of standards will vary with the type of instrument and the length of the light path through the cuvet. The xylose standards up to 0.2 mg./ml. follow Beer's law and seem to be in the range of various photoelectric colorimeters.

Calculations

BLOOD

$$\text{mg. xylose/100 ml. blood} = \frac{A_{test} - A_{blank}}{A_{std} - A_{blank}} \times \text{dilution of blood} \times \text{mg./100 ml. std}$$

URINE

$$\text{mg. xylose/100 ml. urine} = \frac{A_{test}}{A_{std}} \times \text{dilution of urine} \times \text{mg./ml. std}$$

$$\text{g. xylose excreted in 5 hours} = \frac{\text{mg./ml.} \times \text{vol. of 5-hour urine}}{1000}$$

$$\% \text{ xylose excreted} = \frac{\text{g. xylose excreted}}{\text{g. xylose given}} \times 100$$

Oral D(+)-Xylose Test

XYLOSE DOSE OF 25 G.

Subjects selected for the test should have normal renal function as estimated by their ability to concentrate urine, lack of abnormalities in urinalysis, and normal serum urea or nonprotein nitrogen (18).

Withhold food and fluid from subjects at least 8 hours previous to the test. After voiding (discard), have the subject drink a solution containing 25 g. pure xylose dissolved in 250 ml. of tap water. Within 2 hours give another 250 ml. of water to provide adequate urinary flow. Maintain the subject at rest during the 5 hour fast of the test period. Draw a venous sample of blood (clotted or oxalated) after 2 hours. Pool all urine passed for the 5 hours after the xylose has been given. For consistent results the amount of pooled urine specimens should exceed 150 ml. in 5 hours. The reproducibility of the 25 g. dose is excellent (4, 17, 25). Determine the amount of pentose present in both blood and urine specimens according to the method (1) discussed previously.

XYLOSE DOSE OF 5 G.

Occasionally a dose of 25 g. xylose may cause diarrhea and discomfort to the patient so a smaller dose of 5 g. has been recommended (23, 26). This causes no discomfort to the subject and is more economical. The test is carried out similarly to the 25 g. dose. The percentage of recovery is also similar. Less dilution of the urine specimen (1:20) is recommended because less xylose has been ingested.

CHILDREN

Lanzkowsky, Madenlioglu, Wilson, and Lahey (27), recommend a xylose dose of 0.3 g./kg. or a maximum of 5 g.; Benson, Culver, Ragland, Jones, Drummey, and Bougas (18) and Moyer and Womack (28), 0.5 g./lb. body weight up to 25 g.; Clark (29), 15 g./sq. m. body surface area; and O'Brien and Ibbott (6) and Jones and di Sant'Agnese (30), 0.5 g./kg. body weight. The latter report greater reproductibility of results and less diarrhea with the dose of 0.5 g./kg. body weight.

NOTE: The subject is kept at rest during this test as there is evidence (25) that the position of the patient is of importance, particularly in regard to the time at which the peak concentration in blood will occur (3). The slower the solution enters the small intestine the greater the amount which will be absorbed. It is known that the rate of gastric emptying into the small bowel is slower when the patient is in a supine position than when standing or sitting upright.

Discussion

The mechanism of intestinal absorption of xylose will not be discussed as this is beyond the scope of this chapter. Refer to Broitman, Small, Vitale, and Zamcheck (31), Barnett, Goodrich, and Kilgore (32), Christiansen, Kirsner, and Ablaza (4), and Segal and Foley (33).

The rate of excretion is constant, and almost entirely dependent on the glomerular filtration rate. Renal tubular reabsorption is negligible (34). The amount absorbed averages 60% of the ingested dose (4, 25) while 40% appears in the urine during the first 5 hours (35). A much smaller portion is excreted during the ensuing 19 hours (18, 36). After administration of labeled xylose, nearly all of the radioactivity of the urine resides in the unaltered pentose fraction; no evidence of the presence of xylose metabolites is obtained (4). Hence, the measurement of urinary excretion of xylose after the ingestion of a standard dose provides a useful estimate of intestinal absorption, since the rate of excretion of absorbed xylose is high, relative to the rate of utilization.

Absorption of xylose takes place primarily in the duodenum and proximal jejunum while ileal absorption is negligible (37, 38).

The importance of the 2 hour blood level is debatable (3, 39, 40, 41). It is generally agreed to be less valuable than the urinary excretion in diagnosing sprue because of the greater overlap between normals and abnormals. Its chief value is reportedly in ruling out false positives resulting from renal insufficiency, but it also helps in demonstrating incomplete collection of the urine specimen.

It has been recommended that the 1 hour rather than the 2 hour blood level be measured for the following reasons. Overlap between normals and abnormals is much less in 1 hour than in 2 hour specimens (3, 18). While average 1 and 2 hour normal blood levels do not differ greatly, the 1 hour range is narrower. In sprue patients the percentage increase in xylose levels between the first and second hours is large; in aged patients (42) the 1 hour level may be of greater diagnostic significance. The effect of renal insufficiency on blood levels may be less in 1 than in 2 hours.

Normal Values

Adults given 25 g. of xylose excrete in 5 hours an average of 6.5 g. or 26% with a range of 4.1–9.3 g. or a range of 16–36% (analysis of studies made by 13 authors). One hour blood values averaged 45 mg./100 ml. (7 studies) with a range from 29–72 mg./100 ml. Two hour blood values (7 studies) averaged 35 mg./100 ml. with a range from 15–76 mg./100 ml.

Subjects receiving 5 g. xylose showed an average 5 hour excretion of 30% of the ingested dose with a range of 20–40% (1.0–1.6 g.) The amount of xylose present in the blood after 1 hour in 14 cases (M.R.) averaged 15.7 mg./100 ml. with a range from 8–28 mg. This accompanied a 5 hour urine excretion from 1.0 to 1.43 g.

Children under 3 years of age showed a slightly decreased excretion of 22% [Lanzkowsky et al. (5, 27)], although older children approached adult levels.

For normal values found by other workers using a variable xylose dose, consult the original source.

Abnormal Values

DOSE OF 25 G.

Ross and Nugent (43) recommend that the line between normal and abnormal absorption be drawn at 3.5 g./5 hours (14%) and Christiansen, Kirsner, and Ablaza (4) think that there is a definite intestinal impairment in 5 hour excretion values below 4 g. (16%).

DOSE OF 5 G.

Normal blood and low urinary values are indicative of normal absorption with impaired renal function. Normal urine and blood values, however, do not rule out malabsorption (41). A normal excretion of xylose in the presence of steatorrhea is found in malabsorption of pancreatic origin (18, 31). Normal values may be found in uncomplicated hepatic disease with no consistency of abnormal results (4).

The following notes, although incomplete, give an indication of the many and varied instances in which the oral xylose test has been found useful. Decreased xylose excretion is observed in elderly patients without evidence of sprue or renal insufficiency (42, 44, 45, 46). Such subjects reportedly show a normal 1 hour blood xylose response (42). Hypermotility associated with diseased small intestine, e.g., terminal ileitis, decreases xylose absorption in some cases (4). Hypermotility and diarrhea induced by xylose may lead to abnormally low excretion (46). Subjects on high-alcohol low-protein diets exhibit decreased xylose absorption (31). Partial gastrectomy results in borderline or low xylose absorption, while ileum resection is without significant effect (37). Nonfasting patients absorb less xylose (4). The effect of hepatocellular damage on the results of the xylose test has been debated (4, 45, 47, 48). No consistent impairment is found in pernicious anemia (36, 41, 46). Slight improvement in the excretion of patients with gluten-induced enteropathy following diet therapy is usually found (15, 18, 30, 36, 44), although the values may frequently be in the low, or low normal range.

Normal xylose excretion values were obtained in children with cystic fibrosis of the pancreas, ileocolic resection, and ulcerative colitis (30).

This test has been useful in the diagnosis of tropical and nontropical sprue, although the decreased urinary excretion of xylose in untreated sprue patients is similar to results noted in patients with nontropical sprue and among military personnel with malabsorption (4, 15, 25).

The oral xylose test is simple, rapid, and appears to be a most useful screening procedure for identifying idiopathic and secondary types of malabsorption or steatorrhea.

REFERENCES

1. Roe, J. H., and Rice, E. W., Photometric method for determination of free pentoses in animal tissues. *J. Biol. Chem.* 173, 507–512 (1948).
2. Fischer, A. E., and Reiner, M., Pentosuria in children. *Am. J. Diseases Children* 40, 1193–1207 (1930).
3. Roberts, I. G., Beck, I. I., Kallos, J., and Kahn, D. S., D-Xylose blood level;

time curve as index of intestinal absorption. *Can. Med. Assoc. J.* **83**, 112–117 (1960).

4. Christiansen, P. A., Kirsner, J. B., and Ablaza, J., D-Xylose and its use in the diagnosis of malabsorption states. *Am. J. Med.* **27**, 443–453 (1959).

5. Lanzkowsky, P., Lloyd, E. A., and Lahey, M. E., The oral d-xylose test in healthy infants and children. *J. Am. Med. Assoc.* **186**, 517–519 (1963).

6. O'Brien, D., and Ibbott, F. A., "Laboratory Manual of Pediatric Micro and Ultramicro Biochemical Techniques," pp. 240–241. Harper & Row, New York, 1962.

7. Crowley, L. Y., The application of the AutoAnalyzer to the determination of d-xylose excreted in urine. *Tech. Bull. Registry Med. Technologists* **30**, 213–216 (1959).

8. Ashwell, G., Colorimetric analysis of sugars. In "Methods in Enzymology" (S. P. Colowick and N. O. Kaplan, ed.), Vol. 3, pp. 73–92. Academic Press, New York, 1957.

9. Frankel, S., Miscellaneous tests. In Gradwohl's "Clinical Laboratory Methods and Diagnosis" (S. Frankel and S. Reitman, ed.), 6th ed., Vol. 1, pp. 231–232. Mosby, St. Louis, Missouri, 1963.

10. Dische, Z., Spectrophotometric method for the determination of free pentose and pentose in nucleotides. *J. Biol. Chem.* **181**, 379–392 (1949).

11. Bowness, J. M.. Application of the carbazole reaction to the estimation of glucuronic acid and glucose in some acidic polysaccharides and in urine. *Biochem. J.* **67**, 295–300 (1957).

12. Avigad, G., Zelikson, R., and Hestrin, S., Selective determination of sugars, manifesting enediol isomerism by means of reaction with tetrazolium. *Biochem. J.* **80**, 57–61 (1961).

13. Fishberg, E. H., and Friedfeld, L., Excretion of xylose as an index of damaged renal function. *J. Clin. Invest.* **11**, 501–512 (1932).

14. Helmer, O. M., and Fouts, P. J., Gastro-intestinal studies. VII. Excretion of xylose in pernicious anemia. *J. Clin. Invest.* **16**, 343–349 (1937).

15. Gardner, F. H., and Perez-Santiago, E., Oral absorption tolerance tests in tropical sprue. *Arch. Internal Med.* **98**, 467–474 (1956).

16. Fourman, L. P. R., Absorption of xylose in steatorrhea. *Clin. Sci.* **6**, 289–294 (1948).

17. Brien, F. S., Turner, D. A., Watson, E. M., and Geddes, J. H., Studies of carbohydrate and fat absorption from normal and diseased intestine in man. I. Absorption and excretion of d-xylose. *Gastroenterology* **20**, 287–293 (1952).

18. Benson, J. A., Culver, P. J., Ragland, S., Jones, C. M., Drummey, G. D., and Bougas, E., The d-xylose absorption test in malabsorption syndromes. *New Engl. J. Med.* **256**, 335–339 (1957).

19. Sidbury, J. B., The non-glucose melliturias. *Advan. Clin. Chem.* **4**, 40–41 (1961).

20. Reinhold, J. G., Glucose. In "Standard Methods of Clinical Chemistry" (M. Reiner, ed.), Vol. 1, pp. 65–70. Academic Press, New York, 1953.

21. Park, C. R., Reinwein, D., Henderson, M. J., Cadenes, E., and Morgan, H. E., The action of insulin on the transport of glucose through the cell membrane. *Am. J. Med.* **26**, 674–676 (1959).

22. Helmreich, E., and Cori, C. F., The distribution of pentoses between plasma and muscle. *J. Biol. Chem.* **224**, 663–679 (1957).
23. Kerstell, J., Simplified method for the determination of xylose in urine. *Scand. J. Clin. Lab. Invest.* **13**, 637–641 (1961).
24. Larson, H. W., Blatherwick, W. R., Bradshaw, P. J., Ewing, M. E., and Sawyer, S. D., The metabolism of *l*-xylose. *J. Biol. Chem.* **136**, 1–7 (1940).
25. Fordtran, J. S., Soergel, K. H., and Ingelfinger, F. J., Intestinal absorption of D-xylose in man. *New Engl. J. Med.* **267**, 274–279 (1962).
26. Santini, R., Sheehy, T. W., and Martinez de Jesus, J., Xylose tolerance test with five gram dose. *Gastroenterology* **40**, 772–774 (1961).
27. Lanzkowsky, P., Madenlioglu, M., Wilson, J. F., and Lahey, M. E., Oral *d*-xylose in healthy infants and children. *New Engl. J. Med.* **268**, 1441–1444 (1963).
28. Moyer, J. H., and Womack, C. R., Glucose tolerance tests. *Texas J. Med.* **46**, 763–768 (1950).
29. Clark, P. A., The use of the *d*-xylose excretion test in children. *Gut* **3**, 333–335 (1962).
30. Jones, W. O., and di Sant' Agnese, P. A., Laboratory aids in the diagnosis of malabsorption in pediatrics. II. Xylose absorption test. *J. Pediat.* **62**, 50–56 (1963).
31. Broitman, S. A., Small, M. D., Vitale, J. J., and Zamcheck, N., Intestinal absorption and urinary excretion of xylose in rats fed reduced protein and thyroxine or alcohol. *Gastroenterology* **41**, 24–28 (1961).
32. Barnett W. O., Goodrich, J. K., and Kilgore, T. L., Jr., Intestinal absorption. Experimental fat and D-xylose studies. *Am. J. Surgery* **105**, 73–79 (1963).
33. Segal, S., and Foley, J. P., The metabolic fate of C¹⁴ labeled pentoses in man. *J. Clin. Invest.* **38**, 407–413 (1959).
34. Crane, R. K., Intestinal absorption of sugars. *Physiol. Rev.* **40**, 789–825 (1960).
35. Wyngaarden, J. B., Segal, S., and Foley, J. B., Physiologic disposition and metabolic fate of infused pentoses in man. *J. Clin. Invest.* **36**, 1395–1407 (1957).
36. Butterworth, C. E., Perez-Santiago, E., Martinez de Jesus, J., and Santini, R., Studies on oral and parenteral administration of D(+)xylose. *New Engl. J. Med.* **261**, 157–164 (1959).
37. Perman, G., Guelberg, R., Reizenstein, P. G., Snellman, B., and Allgen, L. G., A study of absorption patterns in malabsorption syndromes. *Acta Med. Scand.* **168**, 117–126 (1960).
38. Fordtran, J. S., Clodi, P. H., Soergel, K. H., and Inglefinger, F. J., Sugar absorption test with special reference to 3-0-methyl-*d*-glucose and *d*-xylose. *Ann. Internal Med.* **57**, 883–891 (1962).
39. Thaysen, E. H., and Mullertz, S., D-Xylose absorption tolerance tests. *Acta Med. Scand.* **171**, 521–529 (1962).
40. Joske, R. A., and Curnow, D. H., The *d*-xylose absorption test. *Australasian Ann. Med.* **11**, 4–14 (1962).
41. Bezman, A., Kinnear, D. G., and Zamcheck, N., D-Xylose and Kl absorption and serum carotene in pernicious anemia. *J. Lab. Clin. Med.* **53**, 226–232 (1959).

268   MIRIAM REINER AND HELEN L. CHEUNG

42. Drube, H. C., Der Xylosebelastungstest im Alter. *Klin. Wochschr.* **40,** 518–520 (1962).
43. Ross, J. R., and Nugent, F. W., Gluten-induced enteropathy. *Med. Clin. N. Am.* **47,** 417–424 (1963).
44. Finlay, J. M., and Wightman, K. J. R., Xylose tolerance test as measure of intestinal absorption of carbohydrate in sprue. *Ann. Internal Med.* **49,** 1332–1347 (1958).
45. Fowler, D., and Cooke, W. T., Diagnostic significance of *d*-xylose excretion test. *Gut* **1,** 67–70 (1960).
46. Vartio, T., *d*-Xylose tolerance test. Studies in various anemia and after gastric resection. *Scand. J. Clin. Lab. Invest.* **14,** 36–43 (1962).
47. Merian, P. A., Contribution to experience with the xylose test. *Helv. Med. Acta* **29,** 579–583 (1962).
48. Shamma'a M. H., and Ghazanfar, S. A. S., D-Xylose test in enteric fever, cirrhosis and malabsorption states. *Brit. Med. J.* **2,** 836–838 (1960).

AUTHOR INDEX

Numbers in parentheses are reference numbers and are included to assist in locating references when authors' names are not mentioned in the text. Numbers in italics refer to pages on which the references are listed.

A

Abele, J. E., 162(6), *168*
Abell, L. L., 79(19,35), 85(35), *87*, *88*
Abelson, N. M., *73*
Ablaza, J., 257(4), 262(4), 263, 264, 265(4), *266*
Abu Haidar, G. A., *73*
Abul-Fadl, M. A. M., 218, *221*
Acs, H., 101(8), *110*
Adler, E., 55, *63*
Affeldt, J. E., 171(14), *194*
Alber, H. K., 116(11), *120*
Albert-Recht, F., 79(14), *87*
Alberty, R. A., 160(3), *167*
Albrink, M. J., *168*
Allen, S. I., 47(13), 49(13), *52*
Allgen, L. G., 263(37), 265(37), *267*
Aluise, V. A., 116(11), *120*
Alvin, R. E., 231(3), *236*
Amador, E., 11(69), *16*
Anast, C. S., 140(8), *142*
Andersch, M. A., 211(11), 219(11), 220, 223(6), *229*
Andersen, D. H., 101(2), *110*
Anderson, A. R., 137(2), *141*
Anderson, C. M., 102(10), 109(10), *110*
Anderson, E. P., 3(19), *14*
Anderson, J. T., 9, *16*
Andrews, B. F., 109(18), *111*
Annino, J. S., 4, 6, *14*
Appelhanz, I., 3(14), *14*
Archibald, R. M., 19(2), 28, *30*
Arias, I. M., *72*

Armstrong, A. R., 211(4), *220*
Arnoud, R., 7(52), *16*
Aschaffenburg, R., 213(16), *220*
Ashwell, G., 257(8), *266*
Assous, E. F., 79(30), 85, 87, 89
Astrup, P., 170, 171(8), 177, 179(8), 180(8), *194*
Austin, J. H., 143(4), 144(4), 151 (18), *156*, 181(58), *197*
Avigad, G., 257(12), *266*
Ayers, J. B., 223(21, 22, 23), *230*
Ayres, G. H., 19(4), *30*

B

Babson, A. L., 9(57), *16*, 85(57), 89, 211(7), 218(7, 19), 220, 221
Baens, G. S., 113(2), *119, 120*
Baker, J., 3(17), *14*, 211(3), 217)3), *220*
Bakewell, W., 43(7), 51(7), *52*
Bakst, H. J., 223(16), *229*
Baldwin, D., 3(15), *14*
Baldwin, D. S., 166(13), *168*
Baldwin, E. de F., 192(83), 193(83), *198*
Barclay, J. A., 43(5), 49(5), 51(30), *52, 53*
Barker, H. G., 51(28), *53*
Barnett, W. O., 263, *267*
Barrowcliff, D. F., 6, *15*
Barta, R. A., 109(19), *111*
Basford, R. E., 154(23), *157*
Bates, G. D., 181(57), *196, 197·*
Bates, R. G., 174(28), *195*
Batra, K. V., 85(62), 89
Baumann, M. L., 193(87), *198*

276

Kritchevsky, D., 79(3), 85(3), *86*
Kruk, E., 10(65), *16*
Kuck, J. A., 116(11), *120*
Kuljian, A. A., 47(16), *52*
Kurahashi, K., 3(19), *14*
Kurlen, R., 57(13), *63*
Kurzweg, G., 85(61), *89*

L

La Du, B. N., 200(7), 201(12), *208, 209*
Lahey, M. E., 257(5), 262, 264(27), *266, 267*
Lambertsen, C. J., 171(15), *194*
Landolt-Börnstein, R., 161(4), 163(4), 164(4), *167*
Lang, C. A., 231(7), *236*
Langan, T. A., 79(29), 85(29), *87*
Lange, K., 182(69), *197*
Lanzkowsky, P., 257(5), 262, 264, *266, 267*
Lapham, L. W., *73*
Larson, H. W., 260(24), *267*
Lasowski, E., 223(14), *229*
Lathe, G. H., 59(20), 62, *64, 72*
Lathrop, D., *74*
Laug, E. P., 170(6), 182(62), *194, 197*
Laurell, C. B., 6(39), *15*
Laurell, S., 6(39), *15*
Lawrence, W., Jr., 49(19), *53*
Layne, E. C., Jr., 121, 127, *129*
Le Breton, E. G., 5(32), *15*, 57(16), 59(16), *64, 72*
Lee, C., 179(48), *196*
Leffler, H. H., 79(25), 83(25), 84(25), *87*
Legge, J. W., 147(14), 148(14), 155(14), *156*
Lehmann, H., 2(7), *13*
Leibman, J., 166(16), *168*
Leitner, M. J., 173(24), 180(52), 189(24), *195, 196*
Lemberg, R., 147(14), 148(14), 155(14), *156*
Lennox, W. G., 179(46), 192(46), *196*
Leonard, J. E., 177, 188(75), *195, 197*
Leplaideur, F., 79(32), *88*

Lerner, F., 219, *221*
Leubner, H., 101(4), 109(21), *110, 111*
Levy, B. B., 79(35), 85(35), *88*
Levy, S. E., 167(18), *168*
Lewin, S., 19(6), *30*
Lewinski, W., 49(18), *53*
Light, A., 179(35), *195*
Lilienthal, J. L., Jr., 181(55,56), *196*
Lindemann, R. D., 166(12), 167(12), *168*
Ling, S. M., 223(12,17), *229, 230*
Linn, R. A., 241(6), *243*
Linnig, F. J., 19(3), 28(3), *30*
Litwack, M., 200(6), *208*
Lloyd, E. A., 257(5), *266*
Lobeck, C. C., 109(19), *111*
Loftis, J. W., 223(25), *230,* 232(12), *234, 236*
Looney, J. M., 223(20), *230*
Love, R. H., 213(15), *220*
Lowe, I. P., 201, *209*
Lowenthal, H., 2(7), *13*
Lowry, O. H., 211, 219(10), *220,* 223, 226(7), *229, 230,* 231(8), *234, 236*
Lubochinsky, B., 245, *256*
Luebner, H., 101(4), 109(21), *110, 111*
Lukasiewicz, D. B., 84, *88, 89*
Lund, E., 118(14), *120*
Lundeen, E., 113(2), 119(18), *119, 120*
Lundsgaard, C., 170, 173(2), *194*
Lundsteen, E., 182(61), 193(61), *197*

M

Maas, A. H. J., 179(39,40), *195, 196*
McCaman, M. W., 199, 201, 204, *209*
McDermott, W. V., Jr., 51(38), *54*
MacDonald, R. P., 66(4), 68(7), 70(7), *71,* 223, 225(28), 226(28), 227(28), 228, *230,* 238, 241(8), *243*
MacFarlane, J. C. W., 102(13), *111*
McGann, C. J., *74*
McGavack, T. H., 79(13), *86, 87*
McGeachin, R. L., 3(13), *14*
MacInnes, D. A., 170, 177, *194*

278

SUBJECT INDEX

A

Absorption, xylose, 257
Acid phosphatase, 211
  stability, 2
Alkaline phosphatase, 211
  stability, 2
Amino acids, free, urine, preservation
  of, 12
Ammonia, blood, 43
  urinary, urea nitrogen and, 245
Autoanalyzer, automatic chemical analysis, 31
Automatic chemical analysis
  calculations, 40
  colorimetry, 37
  dialysis and filtration, 35
  general description, 32
  general precautions, 38
  mixing, 34
  multiple analytic systems, 41
  other endpoint measurements, 38
  other processes, 38
  principles of, 31–42
  proportioning, 33
  sample integrity, 33
  sampling methods, 39
  time and temperature control, 36
Azobilirubin absorptivity, 61, 62

B

Berthelot reaction, 245
Bilirubin, direct-reacting, 58, 68
  modified Jendrassik and Grof pro-
    cedure, 58
  modified Malloy and Evelyn pro-
    cedure, 68
Bilirubin (modified Jendrassik and
  Grof), 55–64
  azobilirubin molar absorptivity, 61,
    62

comparison with other methods, 62
  normal values, 63
  principles, 55
  procedure for total bilirubin, 57
  standardization, 58
Bilirubin (modified Malloy and
  Evelyn), 65–74
  advantages of, 70
  azobilirubin, 66
  direct-reacting, 68
  disadvantages of, 69
  normal values, 68
  principle of van den Bergh reac-
    tion, 65
  ultramicro procedure for total bili-
    rubin, 67
Bilirubin standard, recommendation on
  a uniform, 75–78
  acceptable bilirubin, molar absorp-
    tivity, 75
  calibration using preserved standards,
    77
  packaging and preservation, 77
  standard solutions for the assay of
    serum bilirubin, 76
Blood, *see also* Serum
  ammonia, 43
  collection and preservation of, 1
  glucose, 113
  lead, 121
  pH and P_{CO_2}, 169
  urea nitrogen, 245
  xylose, 257
Blood ammonia, 43–54
  apparatus, 44
  methods for measurement, 43
  normal levels, 49
  obtaining and storing specimens for,
    47
  principle, 44
  procedure, 47
  reagents, 46

283